HIGHWATING

LUKAS INGOLD & FABIO TAMMARO

2011

Springer Wien New York

RIEAch Book Series
Editor: Guy Lafranchi

HIGHWAYING
RIEAch (Research Institute for Experimental Architecture)
Berne, Switzerland, Europe

© 2010 Springer-Verlag/Wien
Printed in Austria

SpringerWienNewYork is a part of
Springer Science+Business Media
springer.at

The publisher and editor kindly wish to inform you that in some cases, despite efforts to do so, the obtaining of copyright permissions and usage of excerpts of text is not always successful.

Layout and Cover Design: H1reber, büro destruct, Berne, 2010
Plans, Graphics, Models, Photography and 3D-Renderings by Lukas Ingold and Fabio Tammaro

© pictures
p. 15–17: Google Inc.
p. 86: Schünke Michael, Schulte Erik, Schumacher Udo, Voll Markus, Wesker Karl: Prometheus Kopf und Neuroanatomie, Georg Thieme Verlag, Stuttgart 2006
p. 86: BMW AG

Printing and binding: Holzhausen Druck GmbH
Vienna, Austria

Printed on acid-free and chlorine-free bleached paper
With numerous illustrations

SPIN: 80013902
Library of Congress Control Number: 2010931090

ISSN 1437-7446
ISBN 978-3-7091-0227-5 SpringerWienNewYork

CONTENT

PREFACE

The publication "HighwayING" illustrates the method, the process and the product that have been developed from the thesis project of two graduates, from the Department of Architecture at Berne University of Applied Sciences, Lukas Ingold and Fabio Tammaro. After just three years of study, they were able to handle a complex topic in the short time of eight weeks, producing a remarkable output.

As the focus of RIEAch publications is on experimental architecture and urbanism in connection with conflict situations (political, economic, ecologic, social and cultural), border conditions and regimes of power, "HighwayING" fits perfectly in the RIEAch book series, not only because of the selected topic, but also because it demonstrates a truly experimental approach to the design process that is innovative in terms of both organization and technology. The project raises questions about the future of car technology and mobility and its consequences for society. Furthermore it demonstrates the interaction between education and practice in a unique way without losing the spirit of experimentation.

One consequence of the ever increasing need for mobility, and for systems for the transportation of people and goods, has been the proliferation of highway systems on the outskirts of cities. Over time, these systems have increasingly come into conflict with other urban developments, in the context of which the highway networks, and the noise and pollution associated with them, are seen as scars on the landscape and barriers that block a range of movement flows and urban connections. Frequently they prevent the effective use of land. As a result of this conflict with highway infrastructure development, a strategy has been developed that redefines the interfaces between regional development infrastructure and land use and urban development, placing them in the chain of action of architectural processes.

Advancing to some extent into the field of parametric design, the young architects developed independent processes and strategies to organize and give shape and form to their solution proposals. The traditional clear separation of analysis and design was abandoned in favor of an interdisciplinary model which, on the basis of modeling techniques, enabled a fluid transition between process phases and by means of which it has been possible for holistic solutions to be developed automatically from the conceptual models.
     The topic is introduced by an essay of Per–Johan Dahl (Sweden), architect, teacher and board member of RIEAch, who was involved in condensed design driven research initiatives relating to the interaction between infrastructures (highway) and architecture.

On behalf of RIEAch, I would like to thank the young architects Lukas Ingold and Fabio Tammaro for this great contribution to the RIEAch book series, wishing them all the very best for the future and fervently hoping that they will continue the momentum and go on to produce more projects which make a difference in the way that their first publication "HighwayING" has.

Guy Lafranchi, RIEAch
Berne, 2010

ABSTRACT

With the advent of the automobile in the last century came the construction of highways to link cities with each other. Over time, an extensive highway network emerged and spread across political borders. While, from a large-scale perspective, this appears to enable a seemingly limitless mobility, it has just the opposite effect in the immediate context of a city, as the presence of the infrastructure represents a barrier to urban contexts and movement flows, and fences in the existing urban structures. Since the design of highways is based on purely technical principles, numerous uncovered areas are generated as a byproduct, which prevent an effective land use in these high-traffic locations. Looking forward, the issue of the underlying potential of such idle uncovered areas of infrastructure arises, because dealing with current and future transport technologies represents one of the biggest challenges of urban development.

Professor Guy Lafranchi's experimental design studio "HighwayING", at the Berne University of Applied Sciences in Burgdorf, Switzerland, addressed this issue in the spring semester of 2009, with the objective of developing strategies for making use of uncovered areas and negative spaces of highway systems. This project came about in this context, as thesis work, and deals in an exemplary manner with the highway interchange "Neufeld" in the north of the Swiss capital Berne. This work is an invitation to view the infrastructural heritage of the 20th Century from a new perspective – to discover it anew.

The demonstrated approach consists in graphically capturing both objective and subjective phenomena of this location created by traffic engineers in order to generate context-specific patterns and interpret them. This is done at various scales and is linked to economic, environmental and social aspects. In addition to the patterns generated during data acquisition, the nature of these artificial street landscapes became a topic of the draft, in that the respective geometry determines the structure of the building. The building weaves itself into the looped area and connects the various uncovered areas with bridge-like elements. From the dynamics of the highway intersection, the initially flat building rises up vertically. Due to the continuous connection of all levels with a ramp system in form of a helix, a unique space flow is created, which continues the landscape of the road in the interior of the high-rise building.

The building relates, both in its formulation and in the program, to issues surrounding the automobile and the infrastructure created for it. The building typology developed brings together a hybrid program consisting of a research center and a theme park focusing on man's fascination with all things automotive. Due to the flowing spatial structure, the visitor is taken on a scenographic journey through the individual subject areas of automotive research. In the development of technical-structural elements, such as the ingenious structural system or the movable membrane façade relies on innovative chassis technology, the design process is based on automotive engineering concepts.

Lukas Ingold and Fabio Tammaro
Berne, 2010

THE NEGLECTED SPACES

by Per-Johan Dahl – Malmö, 2010

The current strategies of urban design and development in most European cities are geared towards the densification of existing structures, and to the adaptive reuse of outmoded and centrally located industrial areas. However, despite this instant focus on the historical core, most cities in Europe have been caught up in a state of rapid expansion, which has produced a new kind of urban landscape that rather relates to Dana Cuff's convulsive urbanism than to the neo-traditional visions of concentric growth.[1] Warehouses, terminal buildings, and institutional complexes are mushrooming in the periphery, in conjunction with the proliferation of large residential and commercial districts whose zoning and monolithic structures confirm the continued impact of CIAM urbanism. Indeed, the contemporary European city generally consists of a small, manicured city center surrounded by ever expanding peripheral urbanized spaces, constantly pushing the outer boundaries and blurring the distinction between town and country.

The need to transport growing quantities of people, goods, and raw materials has led to the creation of an extensive infrastructure and a system of highways in the outskirts of cities. These highways, constructed over a period of decades, have gradually been surrounded by residential and commercial districts, generating a new kind of territory that differs radically from historical models. In the populated hinterlands, these highway territories become barriers as well as sources of noise and air pollution. Additionally they often prevent efficient land use, while generating spatial segregation and interruption. The rather late appearance of highways in European city building, their extreme scale and the velocity they promote, their mono-functional legacy, their ability to stimulate paroxysmal development, and their strong physical and psychological impact have together generated a new urban element characterized by a set of problematics that lacks familiar references and normative solutions.

History

Express highway constructions flourished with the progressive urbanization that occurred during the post-World-War-II period, and with the spread of new construction principles. The architectural and urban ideologies of the 1950s and '60s were strongly influenced by the ideas and conditions of the time, including a reliance on automobile transportation, internationalization, and the promotion of social and spatial segregation. These conditions and others provided global legitimacy and a sense of authenticity to the technical means and functional consequences of highway construction. Highways were to become the central feature in the modernist city. From a programmatic and symbolic standpoint, their form would be appropriate to the emerging life styles and the expectations connected with them.

Since the introduction of postmodernism and into the '80s, strong criticism was directed against modernist urbanism. Architectonic ideals and city planning strategies veered heavily towards historical models, along with a combination of greater compactness and more variation in style. The interest in urban densification increased even further during the 1990s, and was associated with the ideal of mixed-use city building. Emphasis upon dense urban cores was promoted, for example by EU initiatives such as the manual for European city planning developed by Richard Rogers.[2] There was considerable discussion about how to build cities from within. Various waterfront projects, such as London's Docklands and Hamburg's HafenCity, were regarded as good examples.

Despite marked interest in compact urbanism, European cities continued to expand in an uncontrolled way, with a jumble of residential, commercial, and industrial complexes, resulting in urban sprawl. While most architectural efforts were focusing on the central sections of cities, the periphery became solely a planning enterprise. Marked by this disciplinary separation, the hinterlands conformed mainly to the scientific tradition of urbanism, while lacking any artistic influences of architecture.[3] Neglected even in these general discussions about urban design and development, superhighway construction in the periphery generated a huge amount of underutilized land, deficient of any identity or character. Additionally, prevailing disciplinary unilaterality made it difficult to deploy critical practices to rethink the role of highway territories in urbanization processes.

The New Frontier

Today, the peripheral landscapes and highway territories offer new and fertile grounds for any architect interested in contemporary urbanism. In these areas, spaces generally have been spared from unproductive layers of nostalgia and sentimentalism, which have been applied to most city centers in Europe and elsewhere since the rise of postmodernism. In these spaces, architects are liberated from the forces of neo-traditionalism and instead can grapple with pure urbanism, framed by the technical and financial developments of the late-twentieth century. Lacking the paralyzing references to historical models, the characteristics of these spaces requires intellectual capacities and design processes that respond to Lebbeus Woods's call for an "open-ended, explanatory nature of the intentions initiating and guiding them."[4] It is in the encounters with extreme scales, velocities, and vastnesses that architects are able to engage in extensive research, and to apply theories and design methodologies that haven't been possible to test in the historical city.

Following Mark Jarzombek's argument, traditional city building models need to be challenged if we are to succeed with the essential task of evolving more sustainable alternatives to common life-spaces.[5] As part of this undertaking, the reconceptualization of highway territories offers a viable alternative to conventional practices. Within the realm of these neglected spaces architecture becomes the new frontier, where we can apply experimentally driven design research to update the disciplinarily underdeveloped practice of highway design. Only when architecture is consulted can highway territories become integral to European city building, and their characteristics investigated, evaluated, and visualized.

1   On the characteristics of convulsive urbanism, see Cuff, Dana: "The Provisional City: Los Angeles Stories of Architecture and Urbanism" (Cambridge, Mass: The MIT Press, 2000).
2   "The Urban Task Force, Towards and Urban Renaissance" (London: Taylor & Francis, 1999).
3   On the disciplinary distinction between architecture and urbanism in this context, see particularly Choay, Françoise: "The Rule and the Model: On the Theory of Architecture and Urbanism" (Cambridge, Mass: The MIT Press, 1997).
4   Woods, Lebbeus: "Experimental Architecture: A Commentary." Avant Garde: A Journal of Theory and Criticism in Architecture and the Arts 2 (Summer 1989): 7.
5   Jarzombek, Mark: "A Green Masterplan Is Still a Masterplan," in Urban Transformation, ed. Ilka and Andreas Ruby (Berlin: Ruby Press, 2008).

SPATIAL PERCEPTION IN INFRASTRUCTURAL CONTEXT

by Lukas Ingold – Berne, 2010

A very long time ago, man has acquired the ability to move faster than his body naturally allows him to do. He managed to expand his radius of activity by riding on animals. The yearning for the mobility that gives us geographical independence[1] is one of the fundamental needs of humankind. In the course of industrialization and the thereby associated technological development one has found new methods of locomotion. Thanks to modern means of transportation like the car, the train, or the airplane, great distances can now be covered in a short time. The speeds made possible by new means of travel have an important influence on the perception of our environment. According to the French philosopher Paul Virilio, who is known as the founder of dromology (from ancient Greek: dromos [race track] and logos [science], "Logic of movement"), the speed of modern locomotion and communication defines our relationship to space and time. Virilio believes that speed annihilates space and compresses time.[2]

The automobile, as the now most prevalent means of transportation, has a significant influence on our space perception. Infrastructures like roads, bridges and tunnels built for individual traffic have created a new level distinct from the natural topography. The speed overburdens humans, especially in regard to their visual space perception ability. Even though a car ride is mainly a one dimensional event, humans can no longer cope with the flood of visual impressions at increased speed. Only distinctive features will still be perceived by them.

Through speed different sensual perceptions are impaired and therefore the driver is decontextualized. The speed theoretician Paul Virilio asks: "Where are we, when we travel? Where is this 'land of speed' which never matches the one we are crossing?"[3] The travellers' relationship with the space they travel through will definitely be disturbed whereby the actual trip loses significance[4]. The trip limits itself to departure and arrival, the passenger becomes a 'human package'[5]. This type of travel is omnipresent in our fast living society.

Because the driver perceives the surroundings with a certain fuzziness, the symbolism of the road environment has become highly significant. With the emergence of the automobile, efforts were made to conform to the requirements of the mobile observer. Besides many novel building types[6] like gasoline stations, drive-in restaurants or motels, advertisements in the form of large scale billboards influenced roadside appearance.

The American city planners Kevin Lynch, Donald Appleyard, and John R. Myer were among the first who focused on the changes of the roadside. They viewed motorists not only as traffic participants, in fact they recognized in them a locked in, partially inattentive audience which is focused only on a limited frame of vision. In an avenue of billboards the mobile public is confronted with a sequence of images and advertising messages and addressed as potential consumers. In their 1964 work "The View From the Road"[7] the three authors compare the roadside to a book that could be read during the ride. A narrative structure should be created for drivers which entertains them during the trip. For the first time the symbolism of roadside advertising became a subject for architectural debate.

In 1972 Robert Venturi, Denise Scott-Brown and Steven Izenour published "Learning from Las Vegas"[8] a sober report about the casino city in the Nevada desert that was created for consumption and entertainment only. For the Venturis the fascination of Las Vegas lies therein that the city had found an independent shape without any assistance from architects or urban planners. Like "flâneurs in automobiles" the Venturis and their students analyzed this cityscape created for the mobile observer.[9]

To a large extent the perception of the city as defined by the Venturis has kept its validity up to today. One appreciates the city as a complex and multi-layered system without consistent rules and assumes that its development can only be planned in so far as it is possible to control it.

In connection with spatial perception in the infrastructural context, references to the motion picture appear often. The speed theoretician Paul Virilio likens a driving car to a film projector. The windshield becomes a projection screen. In an automobile, just like in a cinema, we go on a trip without having to move our bodies. Since the American urban planning theoreticians of the 60s moved the perception of the road into the focus of architectural discourse, the metaphor of film has frequently cropped up. Rem Koolhaas who worked as a movie scriptwriter in his younger years emphasizes that his work as an architect often appears to him as if he were writing a screenplay. "It's about tension, atmosphere, rhythm, about the correct sequence of spatial impressions."[10]

The quote by Rem Koolhaas points out that, if viewed abstractly, it is a phenomenon independent of the medium. No matter if it's films, literature or architecture, it's always about captivating the viewer with a dramaturgy. With a story well told the urban environment can interact with the mobile public that is unable to perceive reality in context due to the speed of its movement.

1 cf. Morisch, Claus: Technikphilosophie bei Paul Virilio: Dromologie, Ergon Verlag, Würzburg 2002, p. 17
2 cf. Moravanszky, Akos: Architekturtheorie im 20. Jahrhundert, Springer Verlag, Wien 2003, p. 360
3 Virilio, Paul: Fahren, fahren, fahren..., Merve Verlag, Berlin 1978, p. 19
4 cf. Schivelbusch, Wolfgang: Geschichte der Eisenbahnreise, Fischer-Taschenbuch-Verlag, Frankfurt am Main 1995, p. 52
5 cf. Morisch, Claus: Technikphilosophie bei Paul Virilio: Dromologie, Ergon Verlag, Würzburg 2002, p. 58
6 cf. Keck, Herbert: Auto und Architektur – Zur Geschichte einer Faszination, Diss., TU Wien, Fakultät Raumplanung und Architektur, Wien 1991, p. 37
7 cf. Appleyard, Donald; Lynch, Kevin; Myer, John R.:The View From the Road, MIT Press, Cambridge (Mass.) 1966
8 cf. Venturi, Robert; Scott Brown, Denise; Izenour, Steven: Learning from Las Vegas, MIT Press, Cambridge (Mass.) 1972
9 cf. Stadler, Hilar; Stierli, Martino; Venturi, Robert; Scott Brown, Denise: Las Vegas Studio: Bilder aus dem Archiv, Scheidegger & Spiess, Zürich 2008, p. 13, 23, 169, 179
10 Rauterberg, Hanno: Worauf wir bauen: Begegnungen mit Architekten, Prestel, München 2008, p. 103: Interview with Rem Koolhaas

ARCHITECTURE AND AUTOMOBILE

by Fabio Tammaro – Berne, 2010

Since the economic recovery after the Second World War almost every Western European and American has been able to own an automobile and so to possess a product of the latest technological standard. Thus the car has come to occupy an important place in our society, be it as a simple means of transportation or as a status symbol.

This affordable means of transport has also quickly become established as an essential consideration in the designs of architects and the visions of urban planners. Besides the necessary transport infrastructure, motorisation has brought with it a range of new types of buildings to cater for and accommodate cars, such as multi-level parking structures and gasoline stations. It was for construction projects such as these that the dynamism, the shape, the design and production principles of the car served as a source of inspiration and a model for architects over and over again.

The most prominent example of the way the car is used as a reference point in architecture is probably provided by Le Corbusier[1]. There are many examples in both literary and architectural works which demonstrate this clearly. Photographs of his buildings, often with a car deliberately placed in the foreground and thus unmistakably connected with the buildings, clearly illustrate the omnipresence of the car in his work.

In his book "Vers une architecture", published in 1923[2] Le Corbusier places a car opposite a Greek temple. In the text he illustrates how, in their relatively short history and in contrast to the tradition of architecture which stretches over thousands of years, car manufacturers have managed to standardize and mass-produce their vehicles. Using a type of building known as the "Maison Citrohan", he demonstrates how mass production similar to that in the automobile industry could be achieved, based on one structural principle and the use of new building materials. To show the influence of the automobile clearly, Le Corbusier names his projects after automobile brands: "Maison Citrohan"[3] (Citroën) and "Plan Voisin"[4]. If nothing else, Le Corbusier sensed opportunities to gain investors from the car industry who would enable him to realize his visions and construct mass-produced buildings.[5]

In "Plan Voisin" the automobile also plays a central role, since it is only this new mode of transport which makes the plan conceivable in the first place. This radical approach to urban planning relies on a strict demarcation of functions and so attempts to meet people's need to be mobile.

Another example of a work influenced by the car is the "Villa Savoye". The shape of this embodiment of modernity offers a striking example of the car's influence. Arrival by car forms an integral part of the design. The floor plan is based on the turning circle of a car. The ramp inside forms a central element of the spatial design It creates a seamless connection between the storeys, by making the floor start to incline and then merge with the storey above. This creates a spatial continuum reminiscent of a never-ending road.

Le Corbusier not only transferred elements of the car into architecture, in 1928 he also designed a car himself - like many other famous architects incidentally[6]. In this way, many times architectural designs found their way into the automobile industry, providing it with new ideas.

The above-named works by Le Corbusier show that in architecture we encounter the automobile in an unusual way, not as a vehicle for overcoming spatial distances, but as a metaphor, an analogy or allegory. At the same time we must recognize that the car should not be confused with a simple machine. As a rule, this machine - possibly the most popular analogy of modernity - has no superfluous components and, for this reason, is often used as a model for reducing buildings to their essentials and dispensing with orna-mentation. On the other hand, like buildings, automobiles are equipped with stylistic elements which are not absolutely necessary from a purely technical viewpoint. For example, cars were fitted with tail fins in the 1950s. These aerodynamic components were inspired by airplane design. It seemed that the qualities of aeroplanes had been transferred to automobiles. In reality, the aerodynamic elements merely served to make cars look fast and were not functional from a technical standpoint[7]. Thus the automobile emphasises its own qualities with the symbolism of its shape.

In spite of the many elements they have in common, automobiles and buildings are two different things. Buildings have a longer life cycle than cars and are more resistant to new technologies. Whereas buildings are bound to a material context, the car is a self-sufficient and mobile form of architecture, carried on a tarmac surface. History shows clearly how architecture and the automobile have consistently influenced and inspired one another.

1   Le Corbusier, whose real name was Charles Édouard Jeanneret-Gris, Swiss-French Architect, *1887, †1965
2   cf. Le Corbusier: Vers une architecture, Crès, Paris, 1923
3   Citroën was the first European car company to use conveyor belts in car production, following the example of Henry Ford
4   The "Societé des Aéroplanes Voisin" was a company which manufactured aeroplanes and automobiles from 1905 to 1938
5   cf. Von Moos S.: L' Esprit Nuveau - Le Corbusier und die Industrie 1920–1925, Ernst & Sohn, Berlin, 1987, p. 259
6   cf. Margolius I.: Automobiles by Architects, Wiley-Academy, Chichester, 2000
7   cf. Braess H.-H. / Seiffert U.: Automobildesign und Technik, Friedr. Vieweg & Sohn Verlag, Wiesbaden, 2007, p. 56

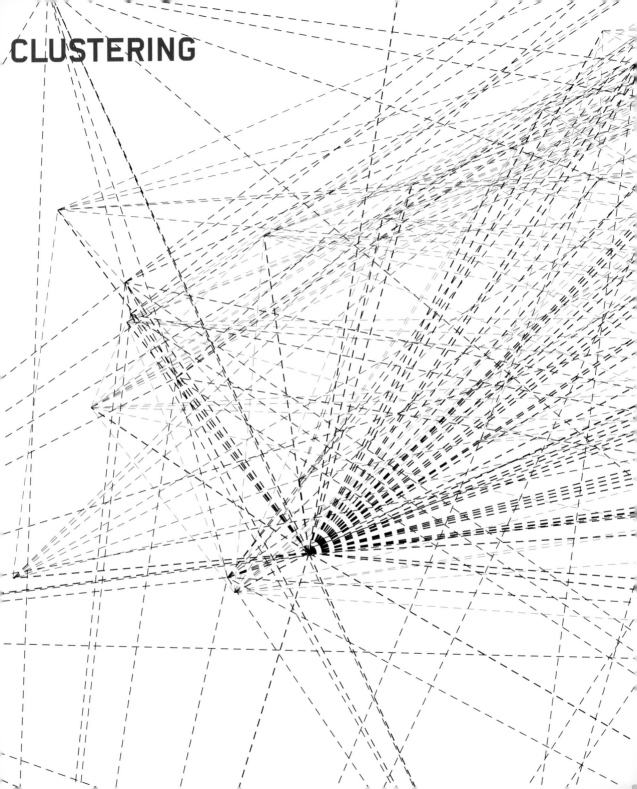

CLUSTERING

Scenario

In terms of both societal and urban planning, the automobile can be seen as one of the most influential developments of the past one hundred years. It has also been a central point of reference in the concepts of architects, urban planners and engineers. Thus new types of infrastructure have been developed around the automobile, and the appearance of our cities has changed accordingly.

Over the past several years, the role of the automobile in the uncontrolled development of our cultural landscape has been increasingly criticized in the architectural discourse. Despite this criticism, in the work-related scenario the assumption is still made that in the future the car will remain a fixture in our daily lives, because when it comes to flexibility, capacity, comfort and convenience the automobile is preferable to other familiar forms of public transportation. According to trends identified by the Federal Roads Office (FEDRO), it is to be expected that traffic on Swiss highways will double over the next 25 years. Considering this development, it can also be expected that the perception and use of the automobile will change in the coming years. According to the prognoses of such leading automobile corporations as Volkswagen, alternative, emissionless engines, advanced electronic driving assistance, and car sharing will be permanent components of the transportation of the future.

It is anticipated that in the near future technological advances will make it possible to resolve problems related to emissions, noise and recycling. The question remains as to what spatial effects our steadily increasing need for mobility will have on our landscapes, as highways are among the forms of infrastructure that offer the best access to our cities, though they also have the potential to cause the greatest disturbances.

Programmatic Context

Express highway construction in Switzerland began in the 1960s. As of today, the majority of the projected national road network has been built. Just a few individual segments remain to be completed. The projected Neufeld site is located on the northern highway section, not far from the city center of the Swiss capital, Berne. The highway spur that provides access to the site is a part of the important Swiss Autobahn A1 East-West route. According to recent statistics, it accommodates an average traffic flow of 98, 500 vehicles per day (source: Automatic Road Traffic Census 2007; Federal Roads Office (FEDRO)).

The highway sections encircling Berne can be seen as a kind of 20th century city wall. Structures reminiscent of city gates can be found at many exits – be it the Westside shopping center which opened in 2008 in Brünnen, or the highway interchange at Wankdorf Stadium which with its many attractive events is a popular point of interest. In contrast to the other aforementioned exits in Berne and despite its close vicinity to the city center, the Neufeld site lacks a clear programmatic definition.

In the last decade, the Neufeld interchange was redesigned numerous times, before finally being completed in 2009 with the construction of an underground highway feeder that relieved the Länggass district of individual traffic. This new connection reinforces the importance of this site for transportation. That in turn provides impetus for developing the unused space.

For many of the residents of Berne and the surrounding communities, the location represents an important throughway and point of reference, although currently it has significance only with regard to infrastructure. The Länggass district with the university is nearby. Education can be seen as a third programmatic priority for Berne in addition to government and tourism. Therefore it would seem to make sense to develop a program for this location in the areas of education, research and development.

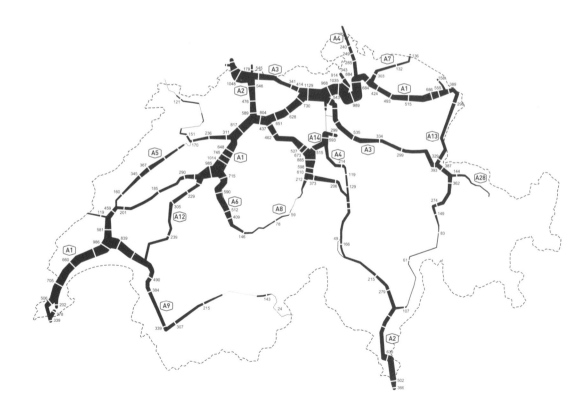

National motorway network average daily traffic (numbers in hundreds)

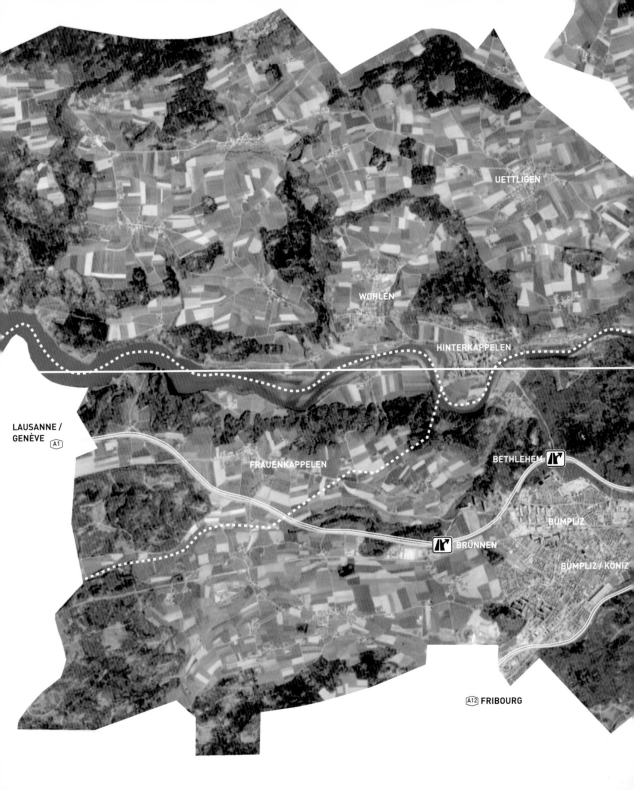

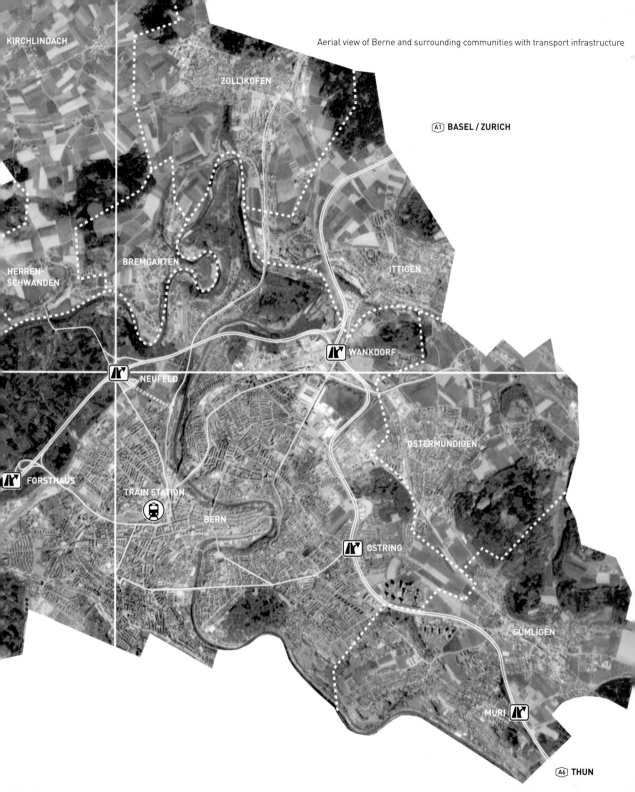

Aerial view of Berne and surrounding communities with transport infrastructure

KIRCHLINDACH

ZOLLIKOFEN

A1 BASEL / ZURICH

BREMGARTEN

ITTIGEN

HERREN-
SCHWANDEN

WANKDORF

NEUFELD

OSTERMUNDIGEN

FORSTHAUS

TRAIN STATION

BERN

OSTRING

GÜMLIGEN

MURI

A6 THUN

Programmatic Context

Primary development by motorized individual traffic and rail (gray)

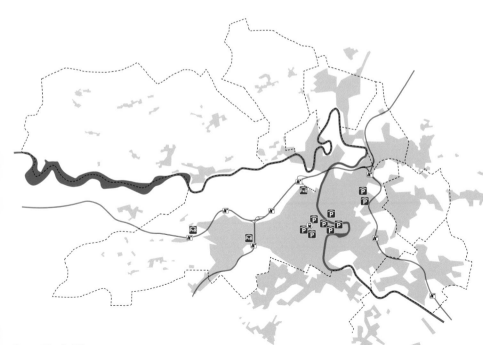

Car parking facilities

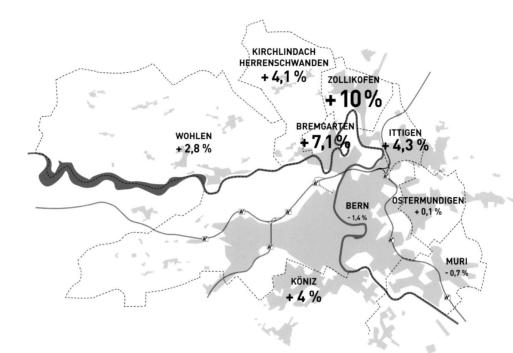

Population development 1999–2009

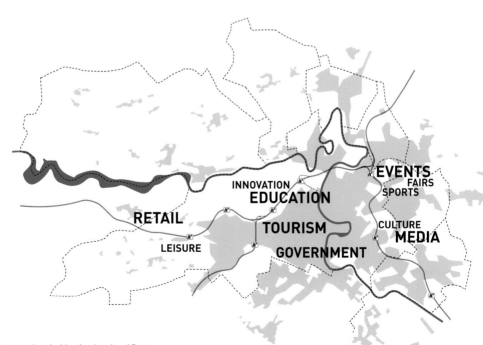

Programmatic priorities for the city of Berne

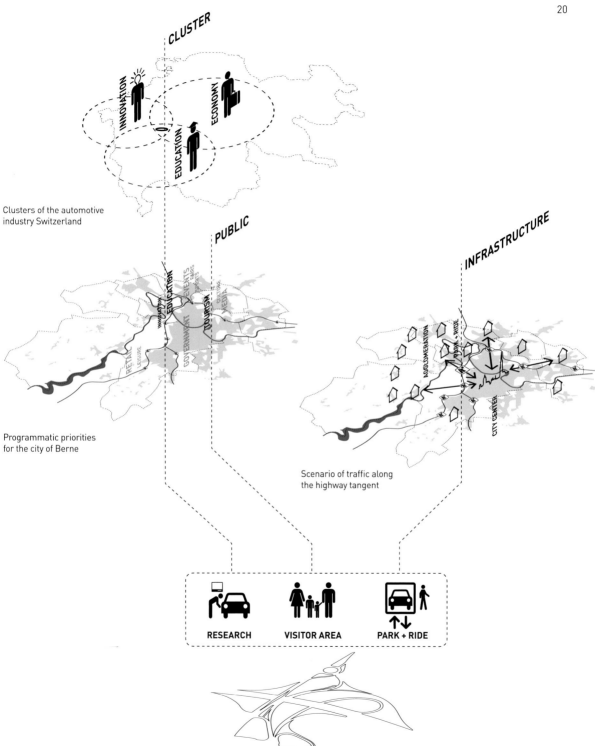

CLUSTER

INNOVATION

ECONOMY

EDUCATION

Clusters of the automotive
industry Switzerland

PUBLIC

Programmatic priorities
for the city of Berne

INFRASTRUCTURE

Scenario of traffic along
the highway tangent

RESEARCH　　**VISITOR AREA**　　**PARK + RIDE**

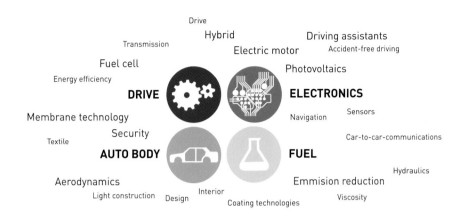

Cluster

Due to the excellent accessibility and the close vicinity to educational institutions, the Bern-Neufeld site offers the ideal basic prerequisites for integration into an economic network, such as the already existing cluster of the Swiss automotive supply industry.

A cluster develops through the merging of research companies, innovative firms and educational institutions. The economic network makes it possible to greatly increase the density of information and communication within a region. Companies are driven to generate innovation by customers, competitors and suppliers, and thus the entire region experiences improvements in the ability to compete and innovate.

The companies involved in the automotive industry can be divided into four research areas: drive systems, electronics, auto body and fuel. With their research capabilities, some of these companies are counted among the most innovative in the industry. Creating a center of professional excellence would make it possible to expand the automotive supply industry's research and development activities in Switzerland.

Program

The research center for transportation technology incorporated into a commercial cluster represents the main feature of the mixed use concept. Another fundamental component is the visitor exhibition area, which offers the public insight into the research activities of companies and schools. Specific topics, such as aerodynamics or drive technology, are explored in interactive displays.

As far as infrastructure in Berne is concerned, there should be an interface between the two overlying traffic systems: the broad national road network and the intricate public transport network. With a Park + Ride as well as a car sharing station, the Bern-Neufeld site will become an ideal transportation hub. These measures follow the trend of relieving the city centers of private automobiles, while at the same time the northern section is better integrated into the public bus and light rail system.

Cluster

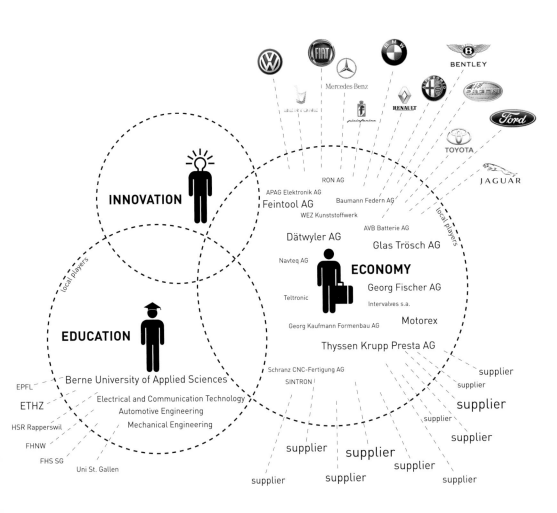

Scheme of automotive suppliers cluster

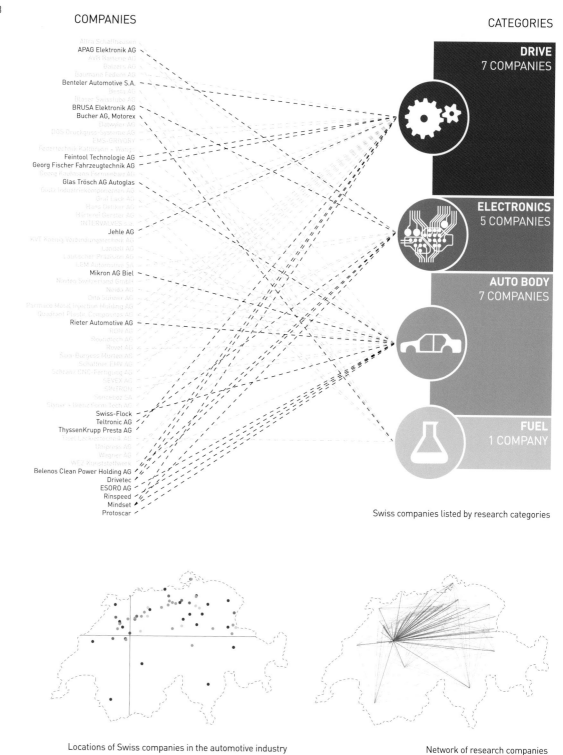

COMPANIES

CATEGORIES

APAG Elektronik AG
Benteler Automotive S.A.
BRUSA Elektronik AG
Bucher AG, Motorex
Feintool Technologie AG
Georg Fischer Fahrzeugtechnik AG
Glas Trösch AG Autoglas
Jehle AG
Mikron AG Biel
Rieter Automotive AG
Swiss-Flock
Teltronic AG
ThyssenKrupp Presta AG
Belenos Clean Power Holding AG
Drivetec
ESORO AG
Rinspeed
Mindset
Protoscar

DRIVE
7 COMPANIES

ELECTRONICS
5 COMPANIES

AUTO BODY
7 COMPANIES

FUEL
1 COMPANY

Swiss companies listed by research categories

Locations of Swiss companies in the automotive industry

Network of research companies

INFRASTRUCTURAL
GEOMETRIES

Topography

The Bern-Neufeld interchange is located in a forest clearing not far from the city center. At this location contrasts clash head on: seemingly undisturbed nature meets geometrically modeled space with curved slopes and sharp dips in the terrain. To ensure optimal traffic flow, the topography of the interchange has been defined by longish on and off ramps. Several modifications of the routes have led to transformation of the plastic soil masses, which are gradually being reclaimed by nature.

Patterns

The unusual conditions required the use of conceptual analysis methods. All the built structures in the area are infrastructure-related – there is no specific architectural environment. The location is characterized by dynamic traffic flow and the resulting emissions. Using parametric methods, subjective phenomena were visualized within the context of location-specific patterns. In order to map the area, a grid was stretched over the area, and the levels of intensity of the phenomena were measured on the intersecting points of the grid. With the help of these parameters, a topographical map of local phenomena was generated using 3D programs. The models generated were then superimposed over the local topography. The distances and levels of activity between the individual landscapes served as a basis of interpretation for further design processes. The selected contour illustration creates associations with the engineering work methods used for this location and can be read in two or three dimensions.

Geometry of the Site

Over the course of time, ideal models for highway interchanges were developed, however, these models – such as is the case for Neufeld intersection – are not always implemented in an unadulterated form. Instead, so-called "spaghetti intersections" emerge, which clearly do not correspond to an ideal basic model and thus yield unique geometric structures when photographed from the air. Between the roadways, which are based on a sequence of additional loops, open areas are separated from the surrounding terrain as the negative spaces of the interchange. Without structural measures, these areas become islands inaccessible to pedestrians.

Terrain relief with sections

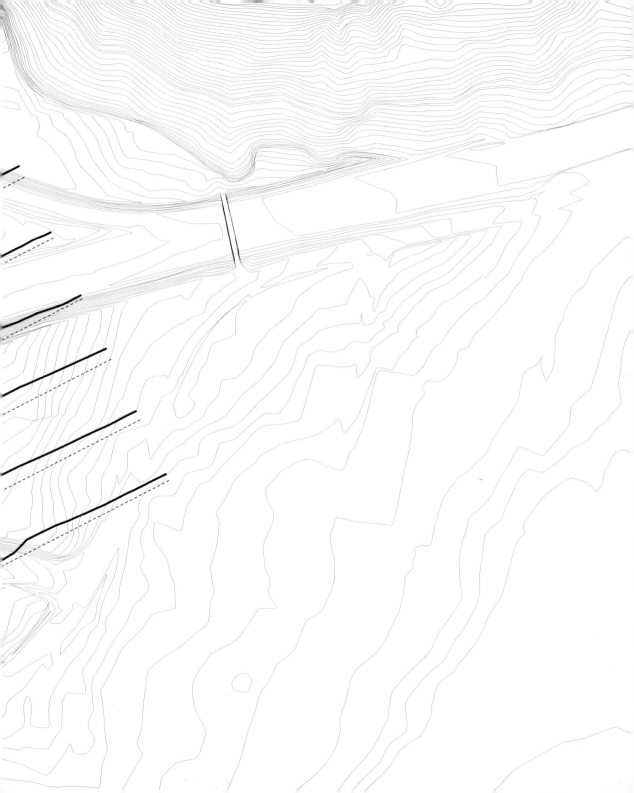

Air pollution

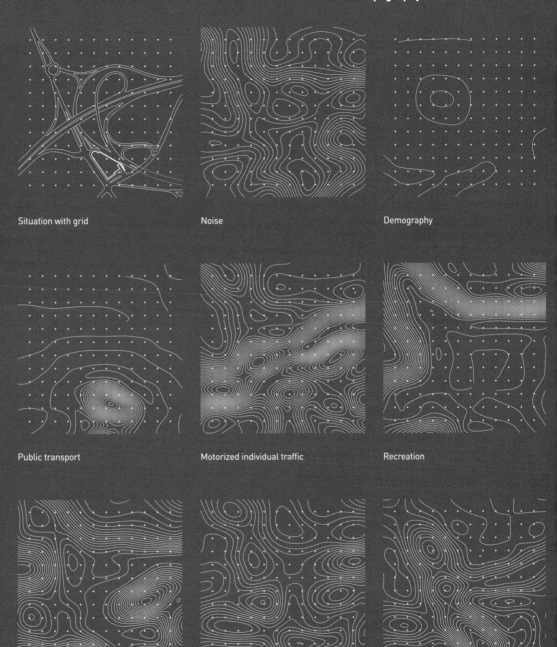

Situation with grid

Noise

Demography

Public transport

Motorized individual traffic

Recreation

Forest

Natural lighting

Artificial lighting

Intensities of the local phenomena in 2D contour line representation

Air pollution

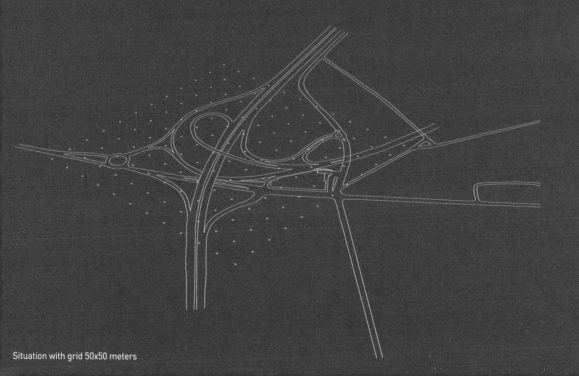

Situation with grid 50x50 meters

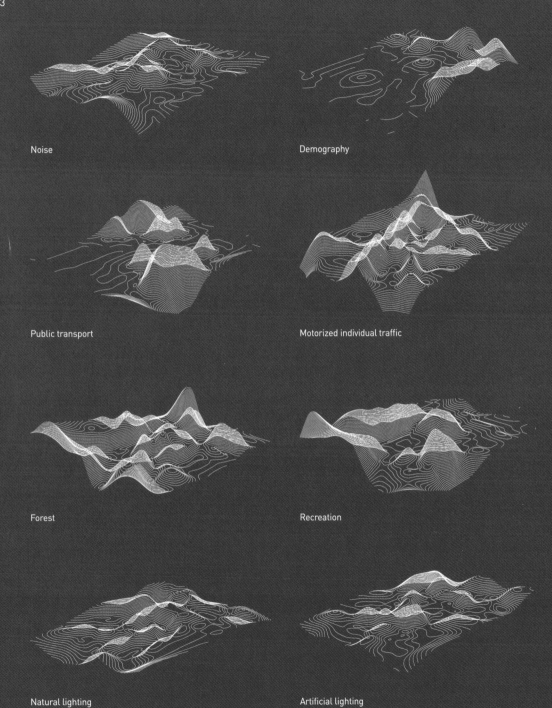

Noise

Demography

Public transport

Motorized individual traffic

Forest

Recreation

Natural lighting

Artificial lighting

Intensities of the local phenomena in 3D contour line representation

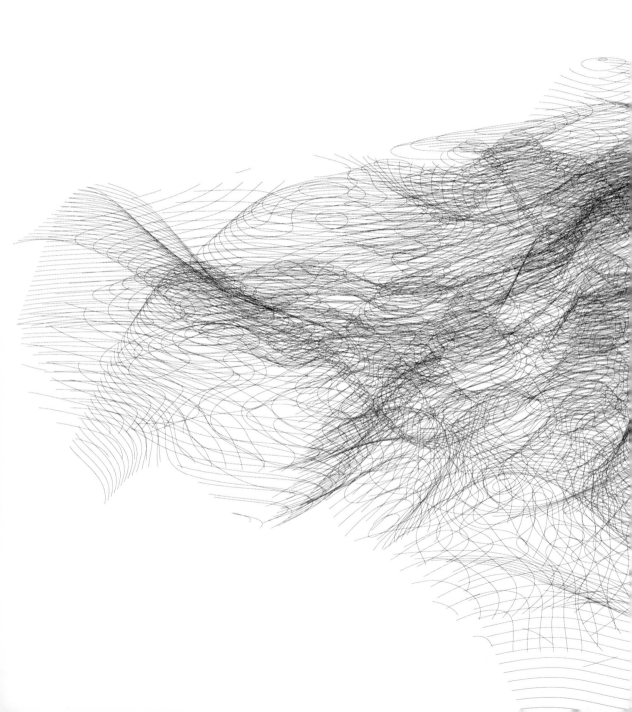

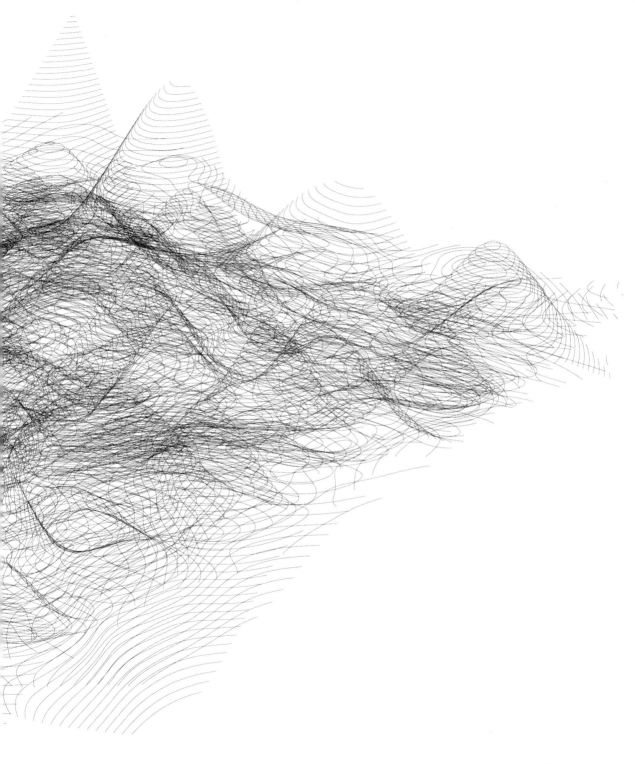

Superposition of all the phenomena 3D

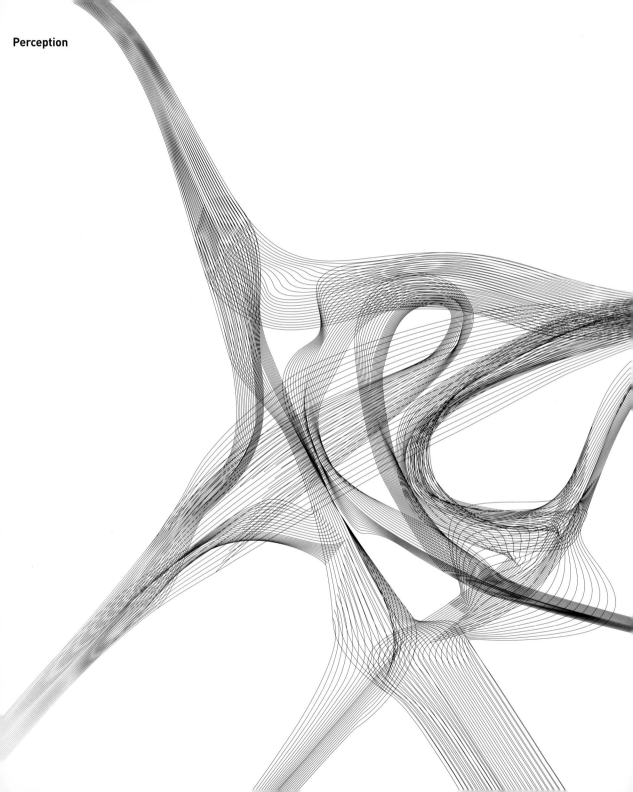

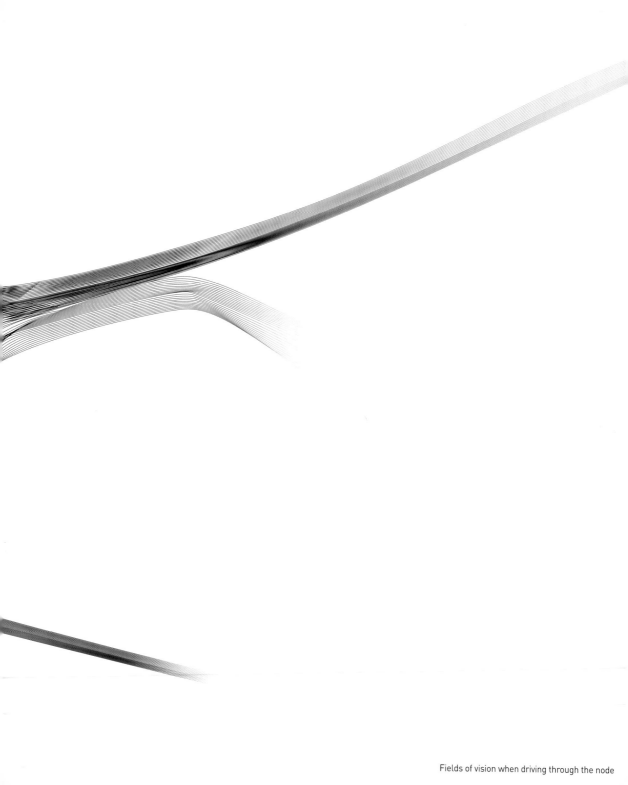

Fields of vision when driving through the node

Geometry of the Site

Interchange at Bern-Neufeld with construction radii and rest areas

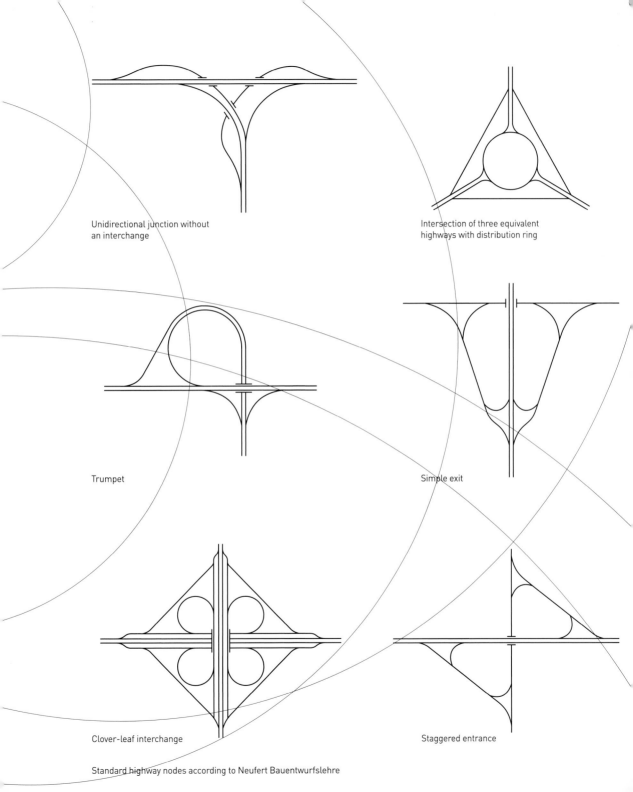

Unidirectional junction without
an interchange

Intersection of three equivalent
highways with distribution ring

Trumpet

Simple exit

Clover-leaf interchange

Staggered entrance

Standard highway nodes according to Neufert Bauentwurfslehre

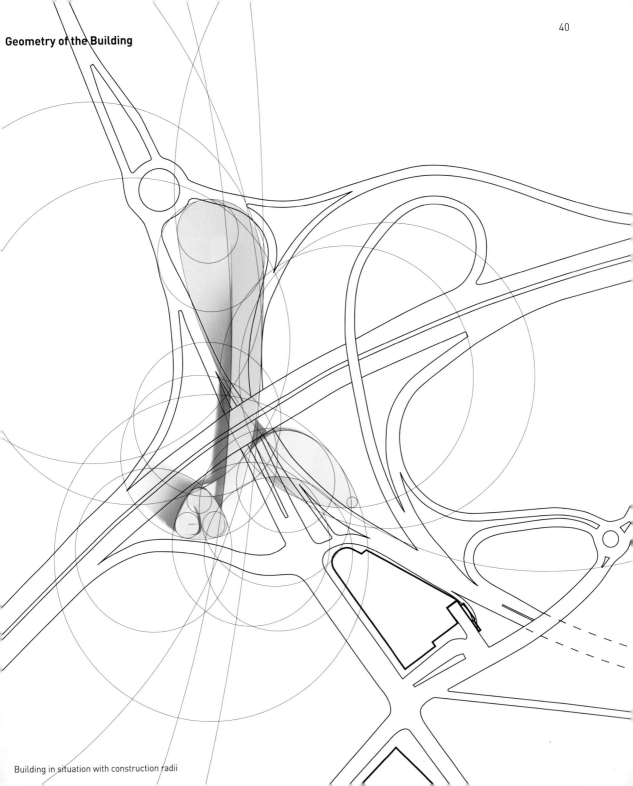

Geometry of the Building

Building in situation with construction radii

Geometry of the Building

The character of the highway landscape, with its extensive loops, is assimilated into a circular structure. The projected building extends over several open spaces and joins them together by means of bridge-like connections. The geometry of this artificial area, which was designed using transportation engineering, is incorporated into the structural organization of the building. A high-rise structure is placed on the western side of the intersection as a deliberate symbol reminiscent of a city gate.

Traffic Routing

Access for motorized private transport is provided via the traffic circle to the north. The bus stop is located centrally on the main street which passes under the highway. Several pedestrian and bike paths make it easy to reach this facility and also provide good access to the forest, which acts as a local recreation area for the neighboring Länggass area community. The previously inaccessible and unused areas of the loops are now accessible.

Applied Program

The plinth part of the building in the northern section of the interchange forms the interface between private and public traffic. The Park + Ride station and bus terminal are located here. Due to the heavy volume of people passing back and forth, commercial space is provided between them, as well as the entrance to the visitor exhibition area. Above the highway, a two-story arm connects this wing with the research tower and accommodates linear spaces, such as the wind tunnel and an array of workshops. The southern part of the building is intended for educational purposes and includes an auditorium, classrooms and administrative offices. This wing of the building is connected with the research tower above main street. In addition to providing internal access to the building, this connecting area also contains a cafeteria, a meeting point for researchers and students.

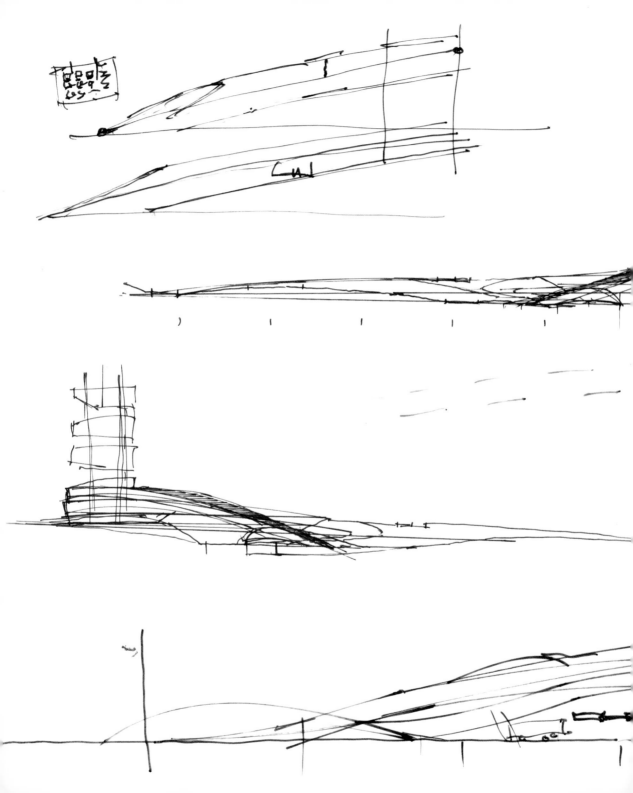

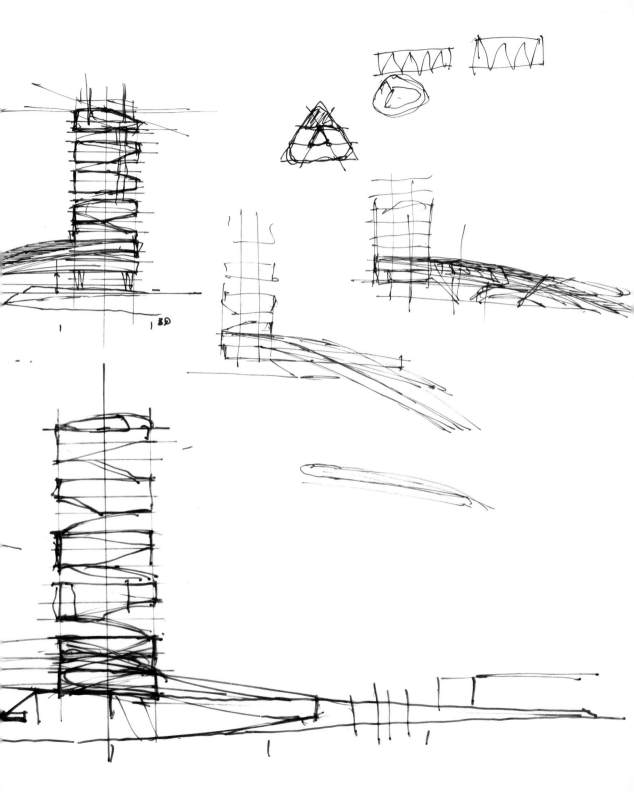

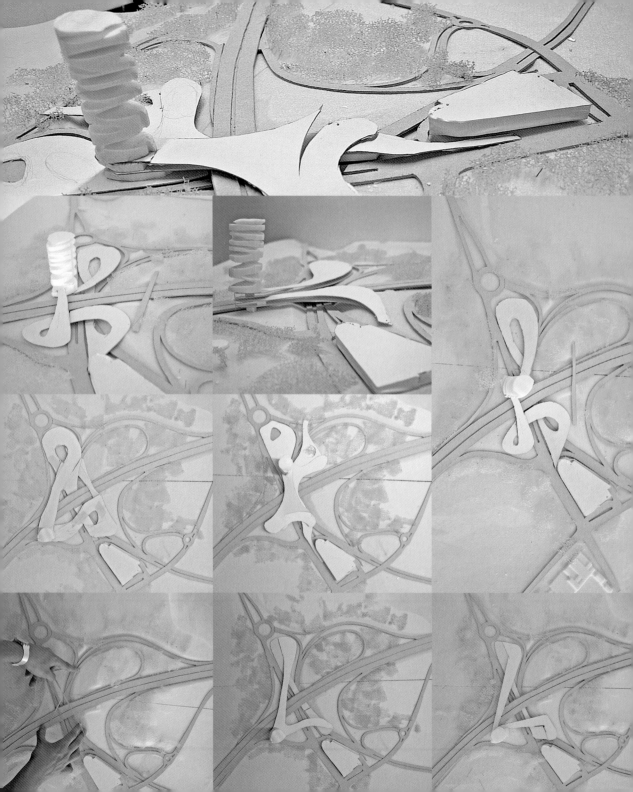

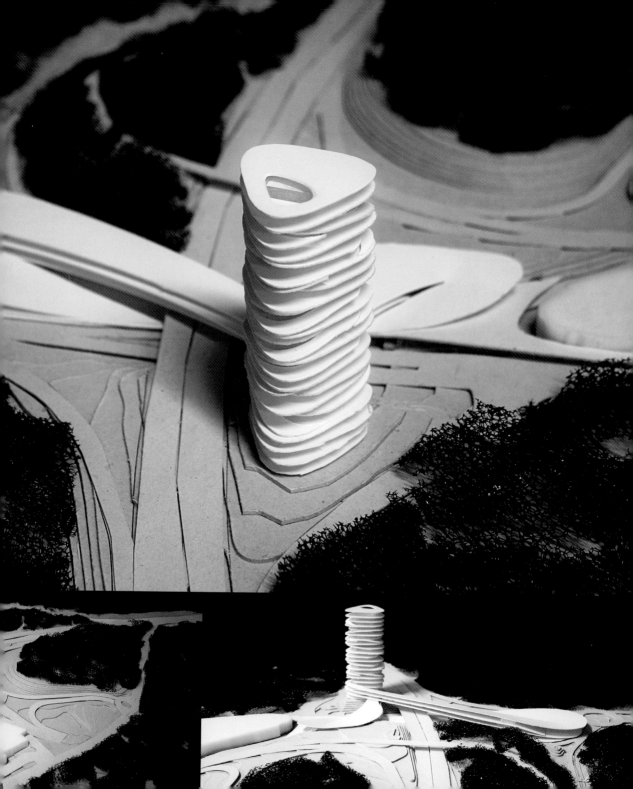

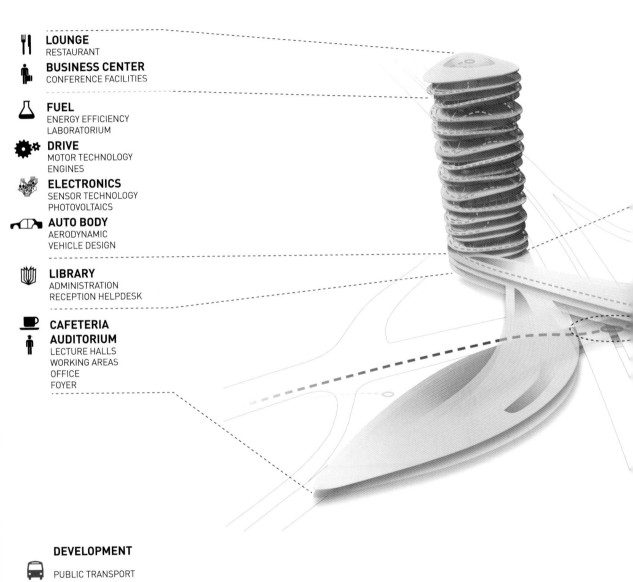

LOUNGE
RESTAURANT

BUSINESS CENTER
CONFERENCE FACILITIES

FUEL
ENERGY EFFICIENCY
LABORATORIUM

DRIVE
MOTOR TECHNOLOGY
ENGINES

ELECTRONICS
SENSOR TECHNOLOGY
PHOTOVOLTAICS

AUTO BODY
AERODYNAMIC
VEHICLE DESIGN

LIBRARY
ADMINISTRATION
RECEPTION HELPDESK

CAFETERIA

AUDITORIUM
LECTURE HALLS
WORKING AREAS
OFFICE
FOYER

DEVELOPMENT

PUBLIC TRANSPORT

MOTORIZED INDIVIDUAL TRANSPORT

VISITOR ROUTE

PEDESTRIAN ACCESS

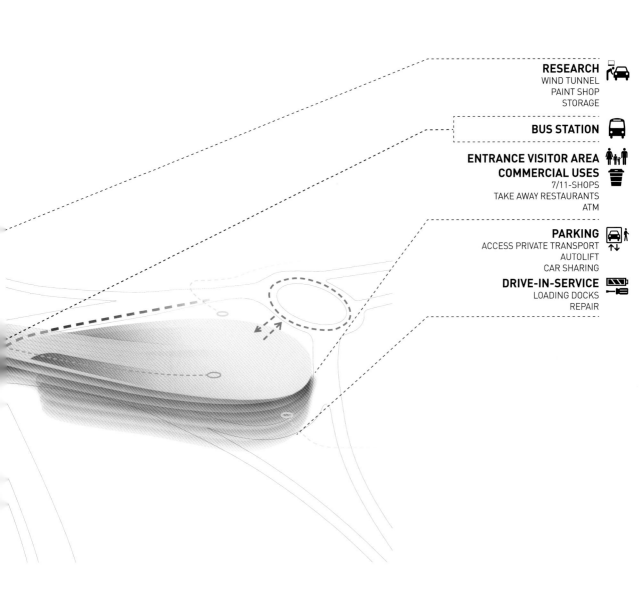

RESEARCH
WIND TUNNEL
PAINT SHOP
STORAGE

BUS STATION

ENTRANCE VISITOR AREA
COMMERCIAL USES
7/11-SHOPS
TAKE AWAY RESTAURANTS
ATM

PARKING
ACCESS PRIVATE TRANSPORT
AUTOLIFT
CAR SHARING

DRIVE-IN-SERVICE
LOADING DOCKS
REPAIR

SEQUENCES

Programmatic Structure

The concept of the building is defined by the planned vertical organization of the four research areas of the automobile cluster. The themes drive systems, electronics, auto body and fuel are each allotted zones between two and four stories tall. Between the research units are individual, raised visitor levels into which theme-related exhibitions have been integrated. A spiral-shaped walkway connects the visitor levels and provides the public glimpses into the research areas. At the base and in the upper stories of the tower there are public and semi-public usage areas, such as a library, restaurant, lounge or conference rooms.

Spatial Structure

Modeled after the structural loops of the highway interchange, the initially horizontal building winds its way up vertically. The spatial structure of the high-rise is based on a sophisticated ramp and deck system that connects all of the floors with one another. Through the vertical street landscape passes a flowing spatial continuum of individual thematic and spatial sequences. The visitor experiences the building on a promenade-like path with numerous changes in direction and varying perspectives and insights.

Static Structure

The design of the static structure is based on the idea of a vertical accumulation of self-contained building components. Within this design, steel space frame rings are stacked on top of one another with a static height spanning several floors. This in turn makes it possible to reduce the number of supporting elements on the visitor levels between the space frame rings. This construction method, loosely based on bridge construction, makes it possible to span large areas and connect individual building components across open areas.

99.00		
	Public	**LOUNGE** BUSINESS CENTER CONFERENCE FACILITIES
80.00	Visitors	ENERGY EFFICIENCY LABORATORIUM
	Research Development	⚗ **FUEL**
66.00	Visitors	MOTOR TECHNOLOGY ENGINES
	Research Development	⚙ **DRIVE**
44.00	Visitors	SENSOR TECHNOLOGY PHOTOVOLTAICS
	Research Development	**ELECTRONICS**
26.00	Visitors	AERODYNAMICS DESIGN
	Research Development	🚗 **AUTO BODY**
		HELPDESK RECEPTION
	Public	ADMINISTRATION **LIBRARY**
00.00		

Programmatic structure

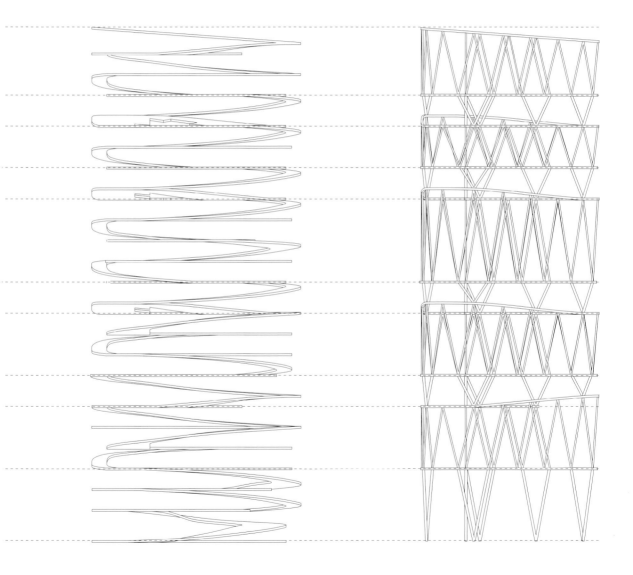

Spatial structure   Static structure

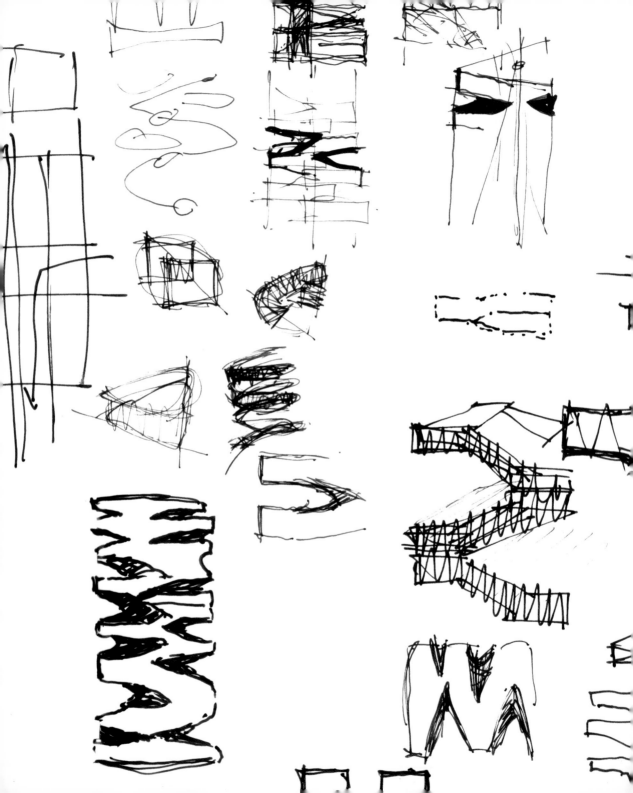

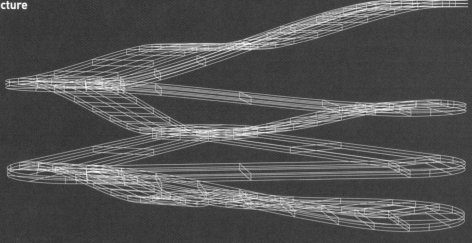

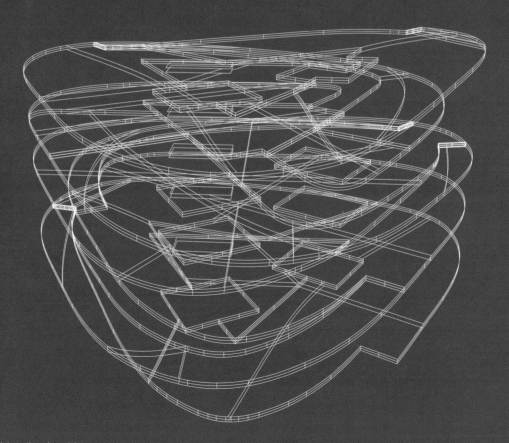

Digital model studies of spatial sequences

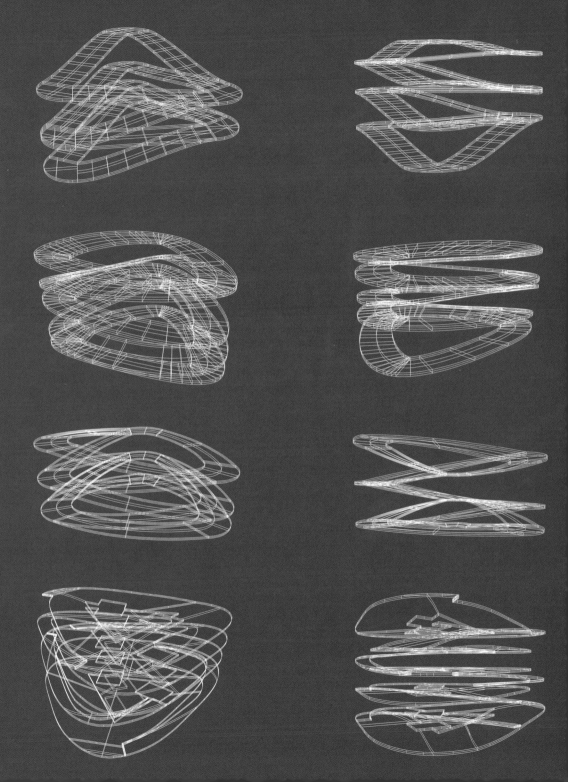

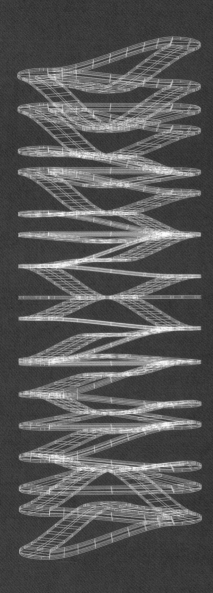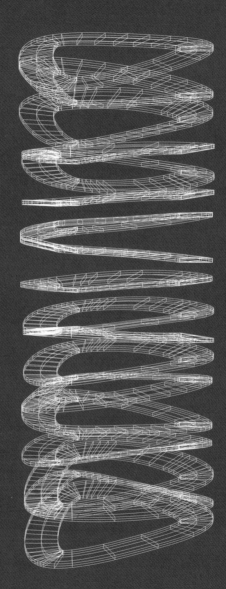

Digital model studies of vertical space flows (various stages of development)

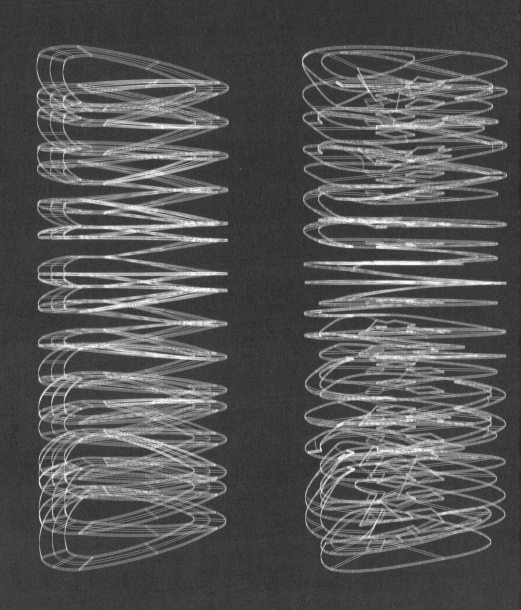

Static Structure

Top view

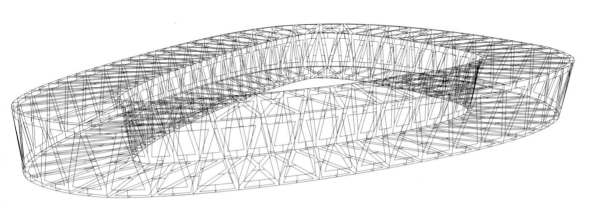

Perspective

Prototype of a single-storey space frame structure

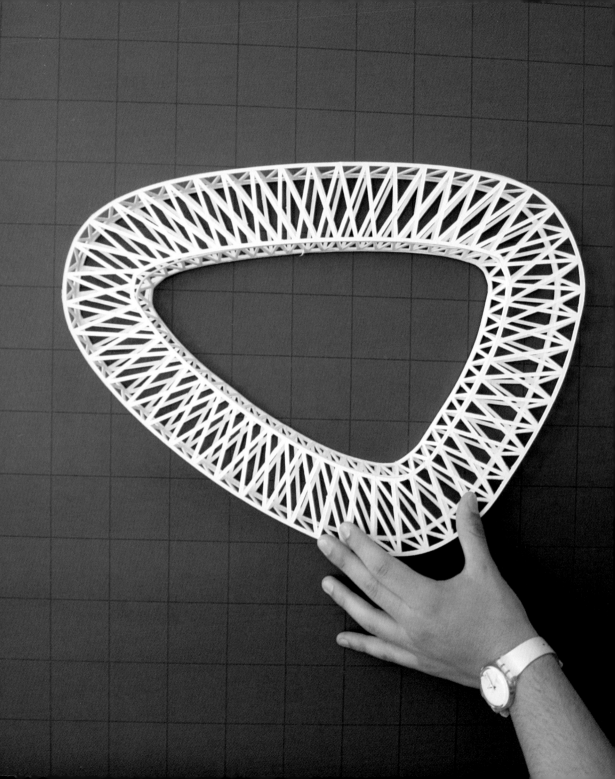

Top view

Perspective

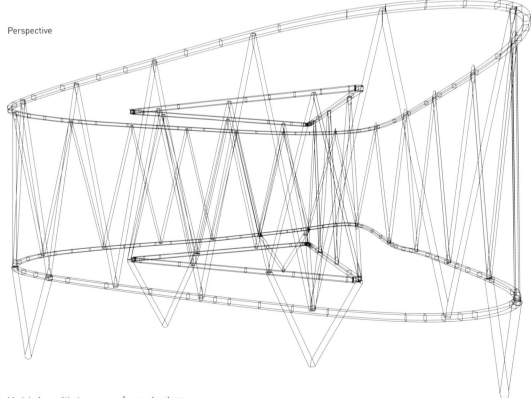

Model of a multi-storey space frame structure

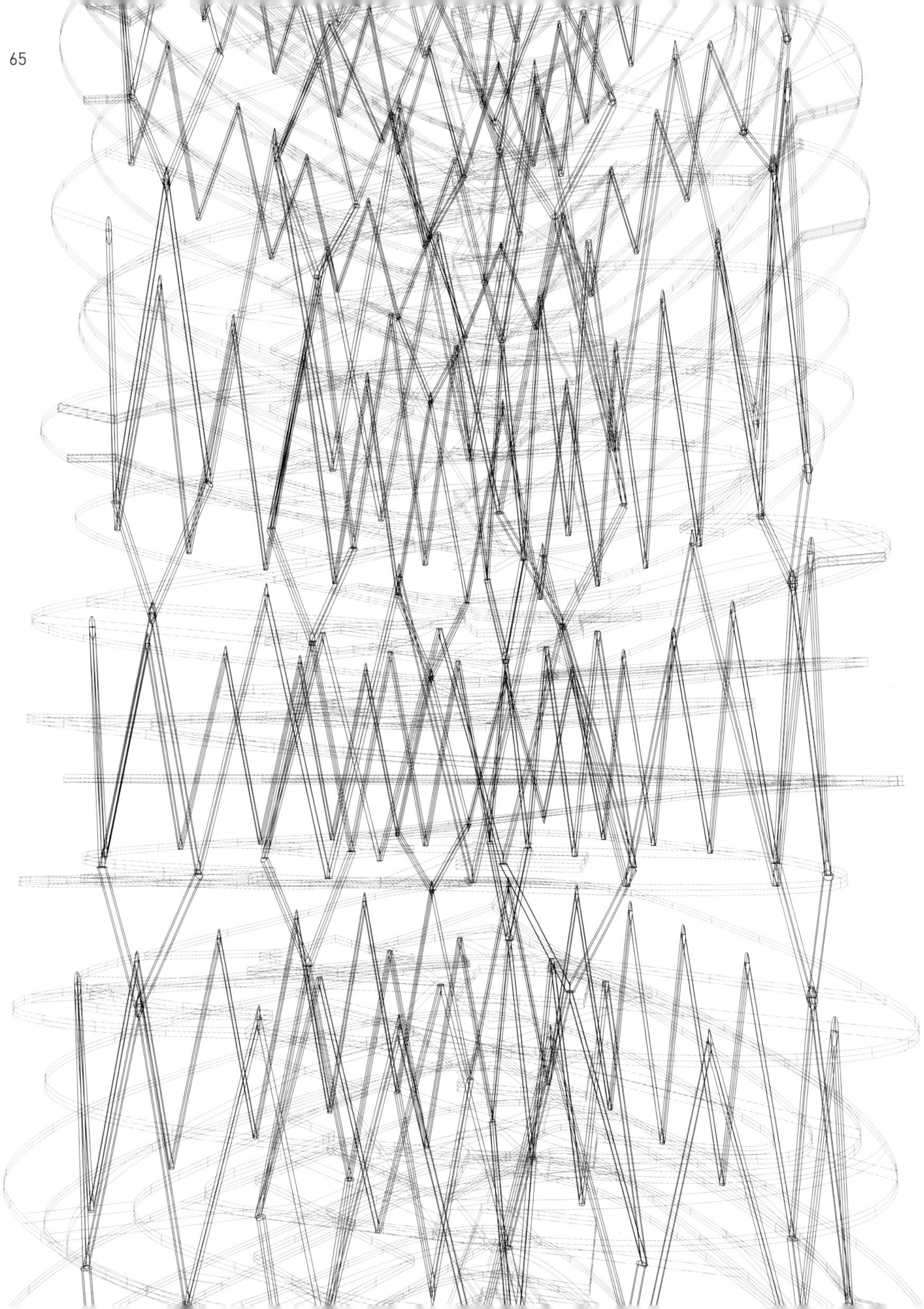

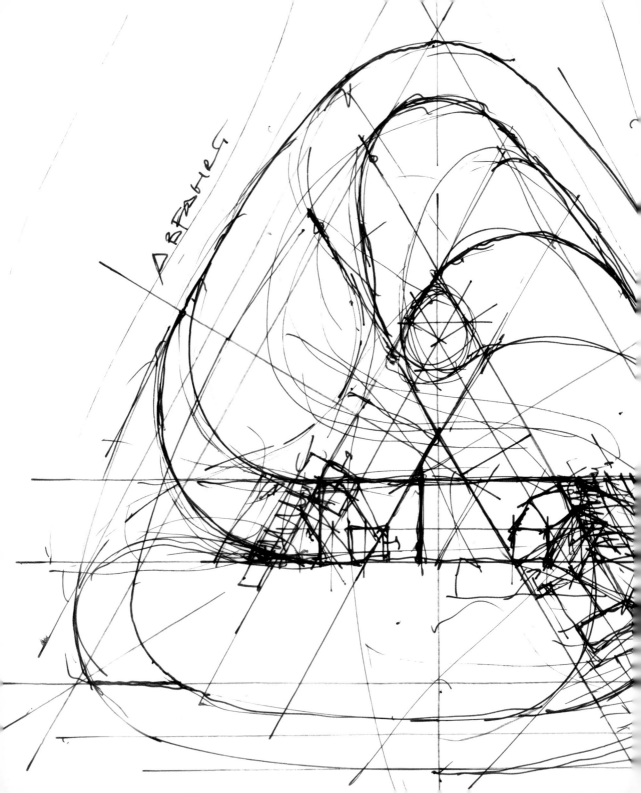

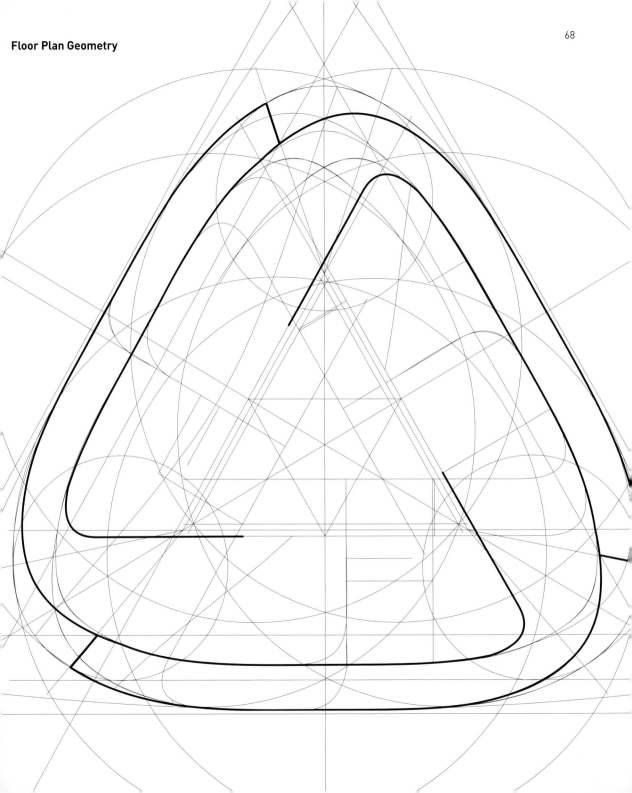

Basic Typologies

Mixed Level (Type I)
clockwise

Mixed Level (Type I)
counterclockwise

Mixed Level (Type II)
clockwise

Mixed Level (Type II)
counterclockwise

Cross Level (Prototype I)

Cross Level (Prototype II)

Visitor Level

Visitor Level
with access ramp

Standard Levels 1:500

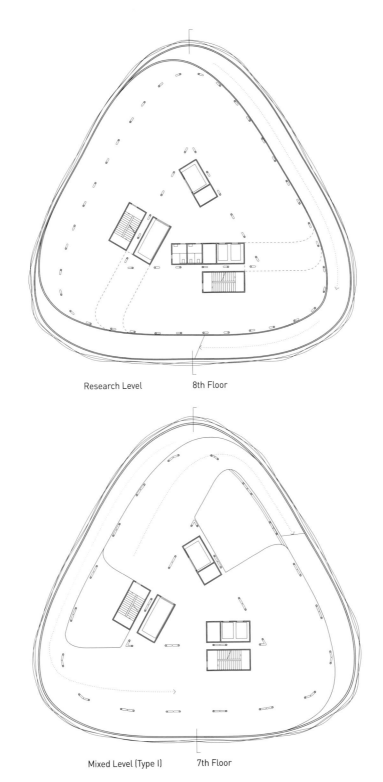

Research Level   8th Floor

Mixed Level (Type I)   7th Floor

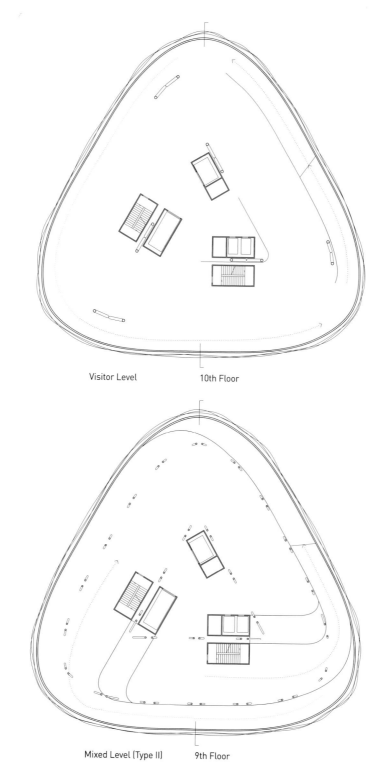

Visitor Level   10th Floor

Mixed Level (Type II)   9th Floor

0   10

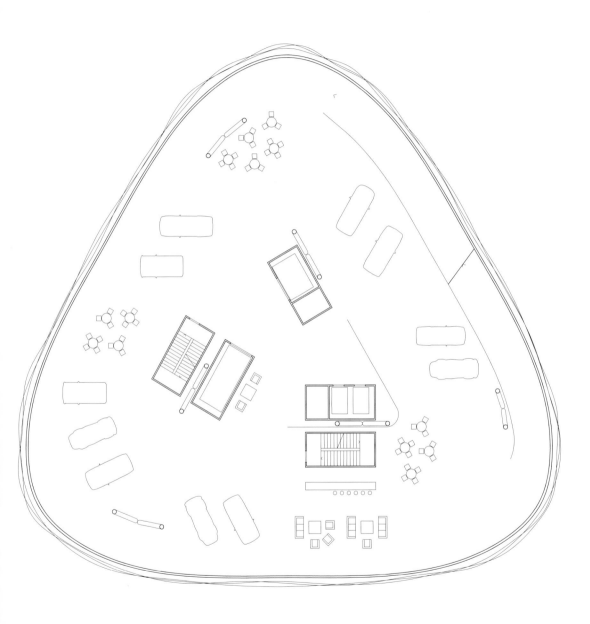

Plan with furniture

0   5

Floor Plan

The floor plan is based on an equilateral, rounded triangle. The surface is then notched and opens like a windmill, which creates gently sloping, wheelchair-accessible ramps and allows for a vertical flow of space. The three utility service shafts support the windmill composition and include an entrance area. Structured around the triangular framework supports, the floor plan is divided into three zones which detach the facade from the load-bearing components.

A freight elevator and several visitor elevators provide vertical access. Pipes and cables are led through two large cable shafts to the upper floors. Emergency exits are located in the two stairwells connected to the building's core units , which also serve to connect the research floors internally.

Scenography

A public walkway leads through the automobile research center and provides many glimpses into the work being carried out by companies and schools. Thanks to the helix-like structure, the visitor experiences a scenographic composition that has quite a dramatic character due to its various theme displays. Visitors can drive a small guest vehicle along the visitor path winding all the way to the top of the tower. Visitors walk or drive the spiral-like path up to the top and use the elevator for the return trip.

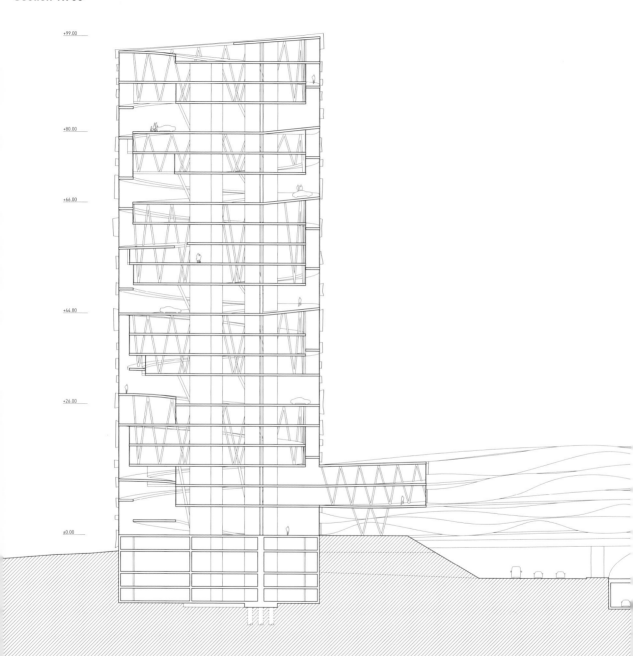

+99.00

+80.00

+66.00

+44.00

+26.00

±0.00

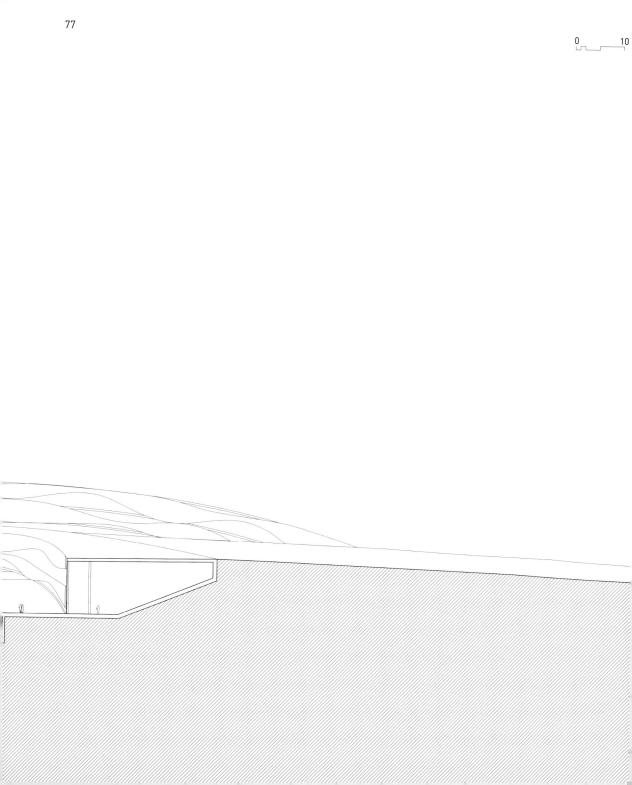

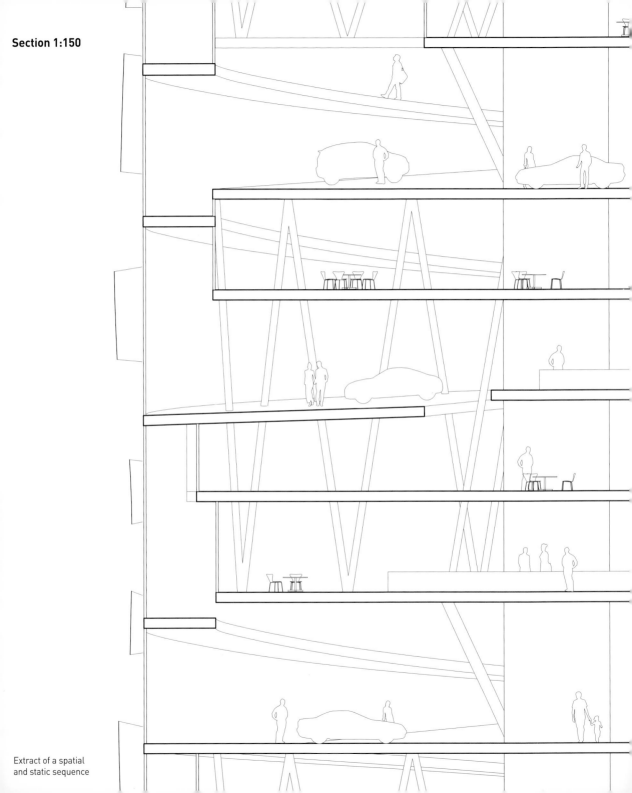

Section 1:150

Extract of a spatial
and static sequence

0   2   5

Scenography

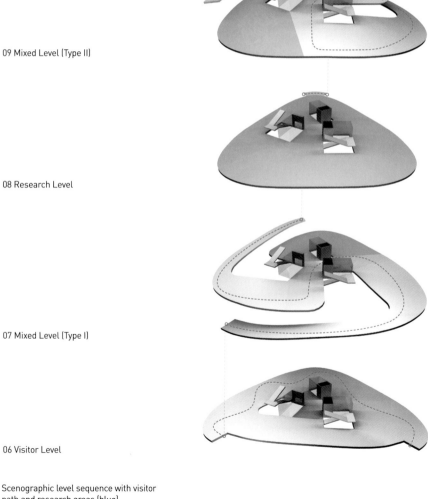

10 Visitor Level

09 Mixed Level (Type II)

08 Research Level

07 Mixed Level (Type I)

06 Visitor Level

Scenographic level sequence with visitor
path and research areas (blue)

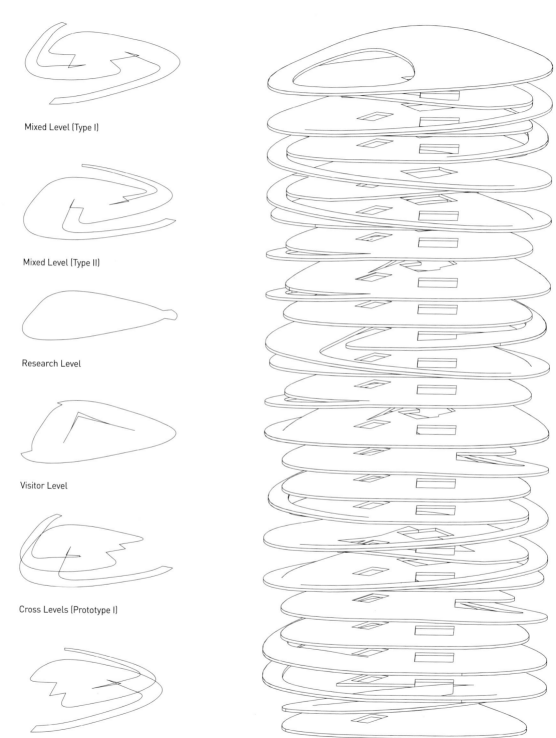

Mixed Level (Type I)

Mixed Level (Type II)

Research Level

Visitor Level

Cross Levels (Prototype I)

Cross Levels (Prototype II)

Levels sequence without static structure

DYNAMIC SKIN

Inspiration

The concept of homogenous, adaptable building shells that can react to outside as well as inside influences and allow for a specific light differentiation in the interior space represents the basis for the development of this innovative facade concept. As there are no relevant examples in the field of architecture, a forward-thinking study from the automobile industry served as the source of inspiration – the BMW GINA Light Vision concept car designed under the leadership of Chief Designer Chris Bangle and presented to the public in prototype form in 2008.

The body of this remarkable automobile does not consist of standard sheet metal but of a durable and extremely expansion resistant textile material that is pulled tightly over a flexible substructure. Visually and structurally, the front and side sections, including the doors, form an interlocking unit. The fabric car has virtually no visible seams or gaps, its shape can be modified as needed, and it almost appears to have human facial expressions – the headlights for example can be opened and closed like a human eye. The facade design represents an attempt to transfer the revolutionary GINA Light Vision system – including the opening of the structure – to the realm of architecture.

Technology

A nearly seamless skin covers the building, determines its changeable shape and responds to the user interactively via sensors. In this way the various potential opening options are controlled and the system can react to external influences, such as solar radiation. With its stretchable, heat and cold resistant, and light-transmitting properties, the material meets today's requirements for an intelligent facade.

It is conceivable that with advances in nanotechnology, photovoltaic cells could also be integrated into the membrane. Thanks to the adaptability of the facade system, the cells can be optimally aligned with the angle of incoming solar radiation, and in this way the outer skin can serve as a long-term, sustainable source of energy.

Expression

With its horizontal layout, in addition to the primary functions of a facade (such as protection from the elements, sun and glare), the exterior skin can be opened in virtually every conceivable manner. The variable degrees of exposure allow specific views to be created as well as different lighting ambiences. The dynamic exterior skin can be made to cast different kinds of shadows, and grants the user a high level of flexibility. With a slight opening, the translucent exterior skin filters the direct light and disperses it softly throughout the room. The refraction generates a diffuse light quality similar to that of a lantern.

The outer expression of the building lends itself to a multitude of compositions, enabling the facade to convey messages. With its great adaptability, the building almost has a human expression, which allows it to interact with its surroundings day or night. The building can become a characteristic landmark in a seemingly endless, anonymous highway landscape and lend the area a new identity.

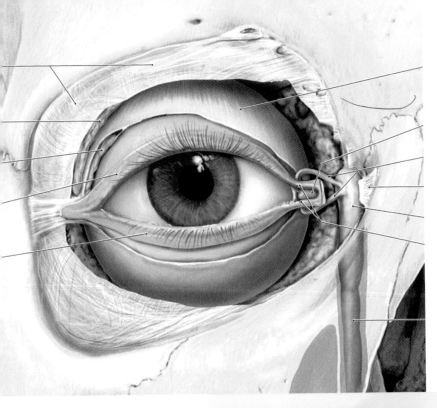

BMW GINA Light Vision

	1 %
	10 %
	20 %
	30 %
	40 %
	50 %
	60 %
	70 %
	80 %
	90 %
	99 %

Opening degree

Facade Studies

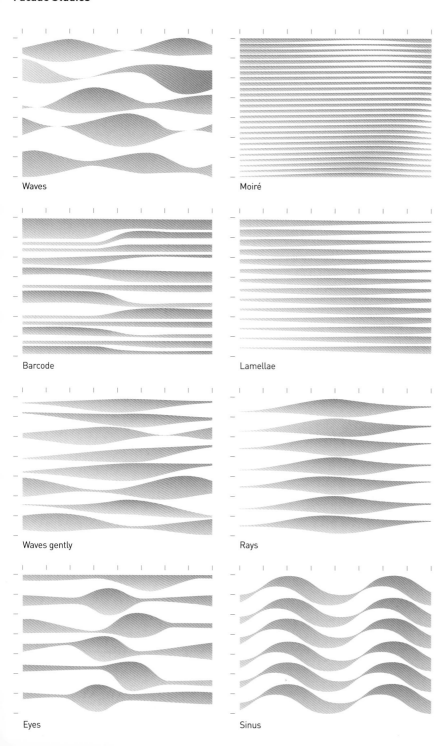

Waves

Moiré

Barcode

Lamellae

Waves gently

Rays

Eyes

Sinus

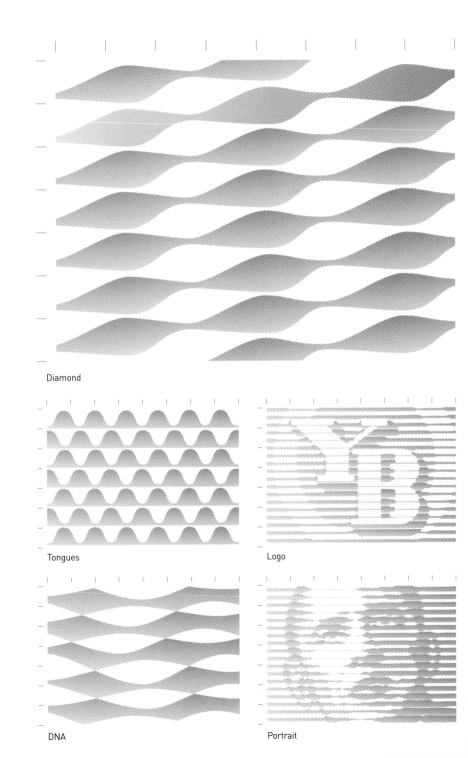

Diamond

Tongues

Logo

DNA

Portrait

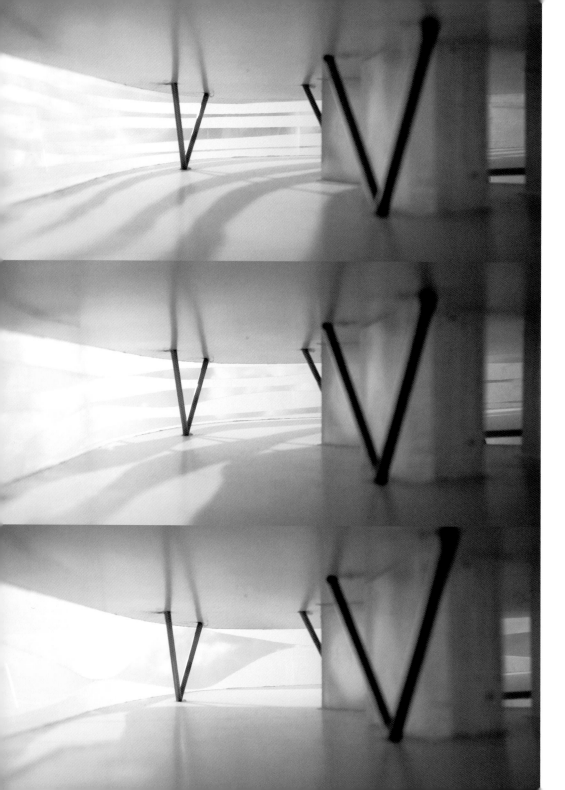

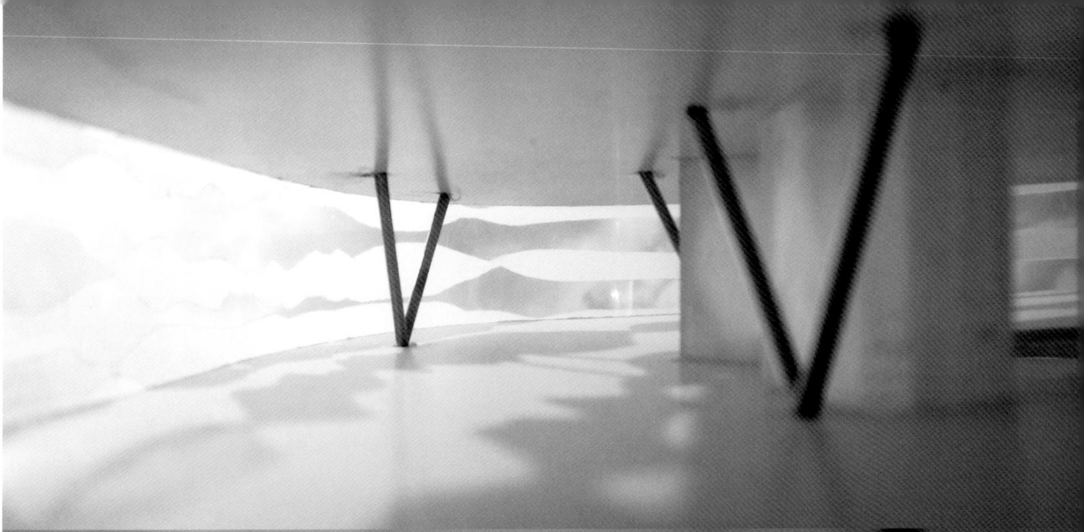

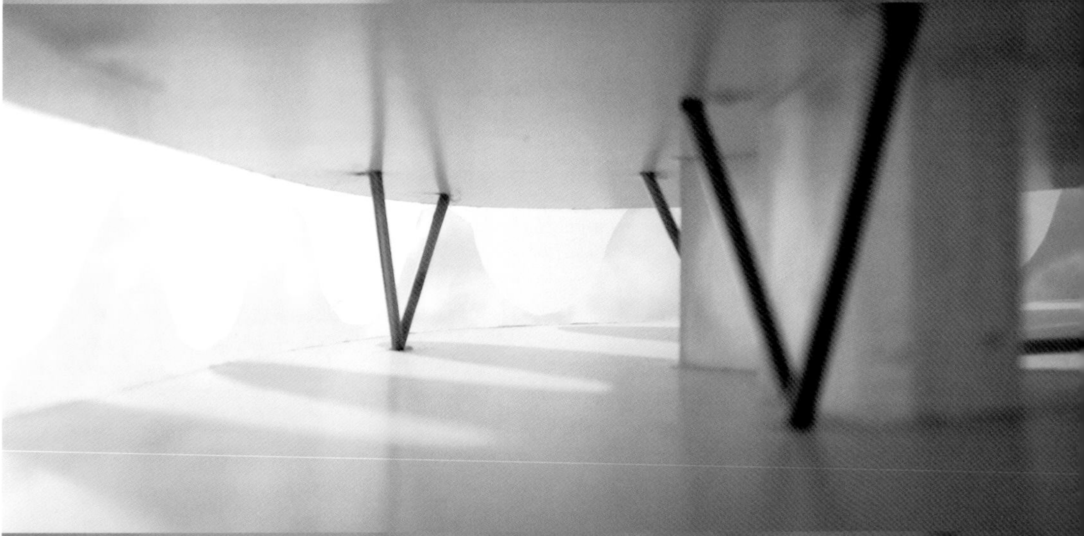

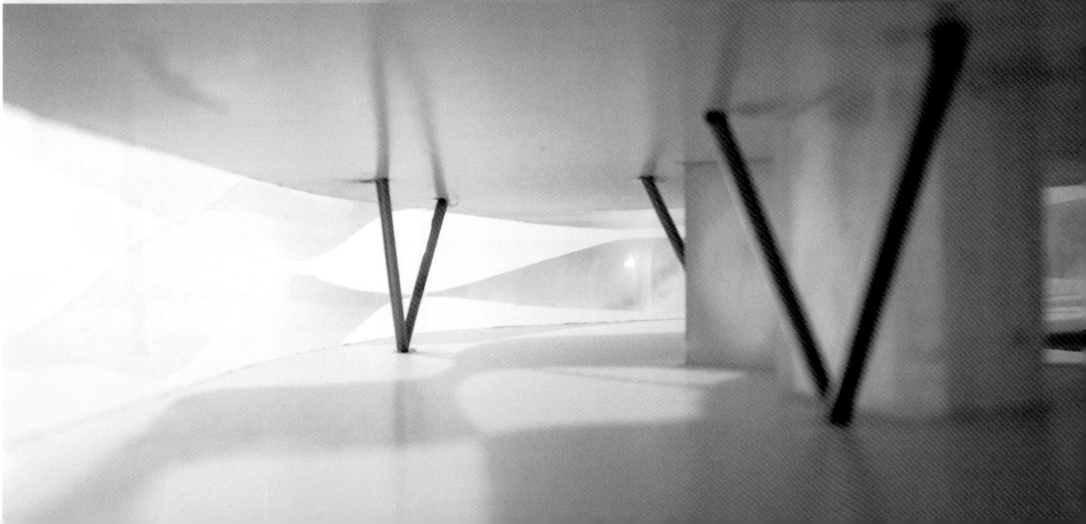

Facade Construction

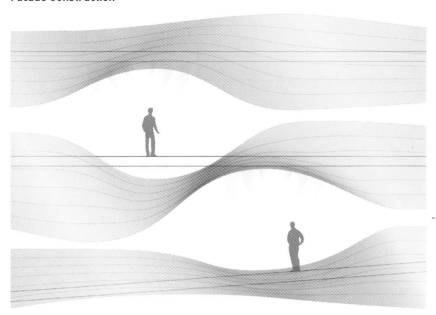

Sensors
Using sensor technology, the facade responds to both internal (visitors) as well as to external influences (sun, views, overall composition. etc.).

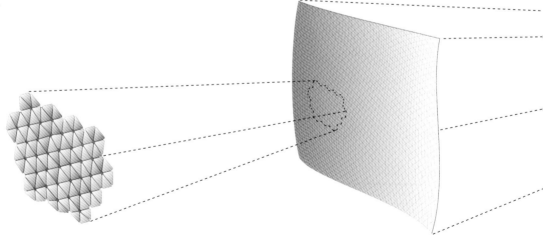

Energy generation
Implementation of photovoltaic cells.

Membrane technology
Highly durable and distension-resistant textile material, stretched over movable base construction.

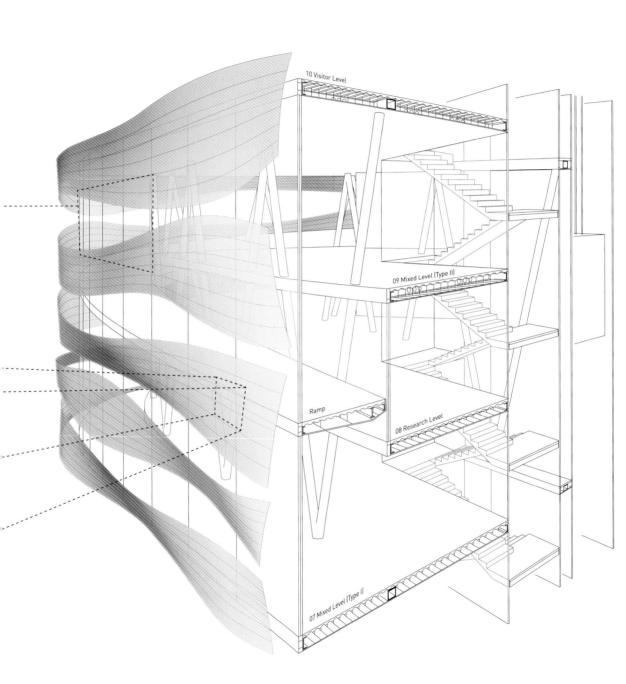

10 Visitor Level

09 Mixed Level (Type II)

Ramp

08 Research Level

07 Mixed Level (Type I)

ACKNOWLEDGMENTS

We would like to thank the Research Institute for Experimental Architecture RIEAch and especially our thesis supervisor, Guy Lafranchi, director RIEAch, for allowing us to have the opportunity to introduce our work to a broad public. For the motivational support, inspiring discussions and ever constructive criticism, we would like to sincerely thank him and the following professors from the Berne University of Applied Sciences: Daniel Boermann (for assistance with construction-related matters), Martin Dietrich (for support with structural design), and Dr. Tim Kammasch (for helping with the scientific editing). In addition, we would like to thank Jacques Wüthrich for his valuable support in designing and building prototypes and for his tireless dedication to the school of architecture in Burgdorf.

The first ideas for this project emerged in the course of the friendly collaboration at the design studio HighwayING (Spring 2009), which turned out to be essential for the design process. While Beat Heuberger worked alongside us as we defined the scenarios and created the plans, the patterns featured in the Infrastructural Geometries chapter were derived in cooperation with David Lüthi, Sanjin Kanesic and Tommy Neuenschwander. The time we spent together was intense, educational and amusing. We would like to sincerely thank our fellow students.

During our study of architecture and work on this book, we were supported by family members and friends in various ways. All those who have helped make this work such a success also deserve very special thanks.

BIOGRAPHIES

Lukas Ingold, born February 27, 1985, in Langenthal (Switzerland) and Fabio Tammaro, born on January 25, 1985, in Biel/Bienne (Switzerland), each graduated from apprenticeships as architectural engineering draftsmen (2000–2004) and then obtained Professional Baccalaureates, Technical Option. They then studied architecture (BA Arch) at the Berne University of Applied Sciences Burgdorf (2006–2009), where they both graduated with honors. As students, both had the opportunity to gain professional experience abroad. Fabio Tammaro worked in an Interior Design studio in Bangkok (2007), while Lukas Ingold traveled to India, under a student exchange program, where he attended the schools CEPT in Ahmedabad and KRVIA in Mumbai, and he participated in a project studio in Bangalore (2008). Currently, both are completing their master's degree (MSc Arch) at the Swiss Federal Institute of Technology Zurich (2009–2012).

www.lukasingold.ch
www.fabiotammaro.ch